METHANOL
And Other Ways
Around the Gas Pump

John Ware Lincoln

Garden Way Publishing
Charlotte, Vermont 05445

To Gisela

Printed in the United States by The Village Press, Inc., Concord, New Hampshire

ISBN   0-88266-051-9 (paperback)
        0-88266-052-7 (cloth)

Designed by David Robinson

Contents

Introduction   1

*The basic human need for mobility. Alcohol:
beneficial fuel additive in the 1930s. Present needs
for alcohol planning. Elimination of pollution.*

**1   The Drinking Automobile:
Another Diet?   6**

*In praise of the versatility of Mr. Otto's engine.
Illuminating gas: fuel of early inventors. Conservation strategies: streamlining, stratified charge, hydrogen, methane, and methanol.*

**2   Methanol: Instrument of
United States Oil Conservation   18**

*Mixing methanol with gasoline. Blends. Performance
improvements. Emissions and miles per gallon. Use
of 100 percent methanol in the future. Objections
to methanol manufacture. Current projects.*

3   Portable Gas Generators   41

*Gasogens: workhorses of European war-time
transport. Simplest, quickest non-oil conversion.
Loss of power. Their role in wooded areas.
Stationary uses, farm uses.*

4   Alternative Engines and Fuels   58

Exotic and expensive ways to avoid using gasoline. Steam cars. The Stirling cycle. Turbines. Electric cars. Fuel cells. Hydrogen and its production. Oklahoma experiments.

5   Past Uses of Synthetic Fuels   86

World War II gasoline savers: alcohol blending, generators, and taxation. Germany: making liquid fuel from coal. Farm alcohol fuel in the United States. Origins of the gasogen: stationary uses, launches, automobiles. Problems of methane use.

6   The Future of Cars (and Fuels)   106

Industry reluctant to face realities of depletion and pollution. Smaller engines. Flywheel power. A new alcohol process. Biomass conversion. The role of industrial lobbies.

7   What's a Driver to Do?   118

The individual dilemma. A long wait for methanol. The drawbacks of gasoline. Methanol conversion. Possibilities for individual action.

Glossary   127

Bibliography   129

Index   133

Acknowledgments

Many people have contributed to this book, since it first took nebulous form, as a series of articles, in 1942, another time of gasoline shortages. For practical reasons, they can be thanked only categorically. More recent and specific contributors include W. J. J. Gordon, who led me to suspect infirm foundations under the fortresses of technology and to ask dumb questions, such as "Why burn gasoline in your car?"; Axel Svedlund, who sent photographs, drawings, and data on his Saab and its charcoal gas generator; Howard Sams & Co., Inc., who allowed me to reprint the alcohol-gasoline carburetor of 1911; and John Wiley & Sons, who permitted me to quote two prophetic paragraphs from *Principles of Motor Fuel Preparation and Application,* by Nash and Howes (1935).

Special thanks are due Ann Moore for meticulous preparation of manuscript; to Cori Forsner for difficult drawings produced from scanty data in a suitcase drafting room; and to George de Kay and Martin Kessler for reading the manuscript and making encouraging and valuable suggestions.

The book would not have materialized without the generous cooperation and stimulation of Dr. Thomas B. Reed of Lincoln Laboratory of the Massachusetts Institute of Technology. He provided me with refreshment of my meagre chemical knowledge with the use of his many publications and photographs. He read and commented on the chapters on methanol, gas generators, and history.

The skill of Roger Griffith, my editor, transformed the chaos of the original manuscript into order, and clarified my lapses into technical jargon.

J. W. L.
November, 1975

Foreword

This book will in very short order come to be considered the standard reference in a field of enormous national importance and even greater confusion. Energy crisis there is not — rather a crisis in petroleum availability and use and a crisis in the extinction of the public interest due to the power of the oil and federal governmental bureaucracies.

Energy needs can and must be made available from sources that:

1. Are renewable
2. Are within our control
3. Can be used without adverse impact on the environment or public safety

This book represents the first — and most welcome — effort to assemble the wealth of support data behind the conclusion that the needs for energy can *certainly* be satisfied in the transportation sector *from* existing resources and technologies — all with salutary impact on the United States economy.

If you do not yet understand the power of the big — whether it is government, industry, university or medical — to exert control over all our lives, you should appreciate the timeliness and courage of this book. Solutions to problems in America have the tendency more and more to be funnelled through large institutions, which, in time, crush out competitors.

In reading this book, you, like I, will be thankful that we still can get access to common sense coupled with professional expertise. As you read through these pages (and between the lines) ask yourselves, Does the author exaggerate the power of the bureaucracies to frustrate the clear interest of the citizens? Finally, take into account that the "Federal Non-nuclear Energy Research and Development Act of 1974" was passed only after the Senate had eliminated as an area of eligible support synthetic energy supplies from agricultural products and wastes which had been passed by the House of Representatives (HR 14892).

In all respects, this is a timely and worthwhile book which will leave you asking How do I do something about this?

Robert Monks
Former Commissioner
Office of Energy Resources
State of Maine

Petroleum —
"Since early exhaustion of the supply is foreseen, it is worth remembering that the pursuit of happiness was not unknown before 1859."

(Columbia Encyclopedia, 1935)

In 1859 Col. Drake struck oil at Oil Creek, Pennsylvania.

Introduction

Man's mobility, when he became erect and abandoned his forelegs for locomotion, was somewhat limited by the modest length of his hind legs. He could sprint briefly, a few hundred yards, after a deer, or he might extend his radius to a dozen miles, by jogging or walking. Early Greek story-tellers, with wish-fulfilling fantasy, invented the *centaur,* combining a horsey speed and range with a human torso and the capacity to enjoy travel.

This need for better personal transportation was achieved, after a gestation of several millennia, in the motor bike and the automobile. To imagine that man will now lay down his new love-objects on demand, only because the price of fuel is going up, or because his money is said to be gravitating to far-off countries, is a folly comparable to National Prohibition of alcoholic beverages.

When President Ford issued his strong corrective recom-mendations in his State-of-the-Union address on Jan. 14, 1975, his most alert opposition could only present equally unpleasant alternatives to the two-dollar-per-barrel import tax, namely, a consumers' gas tax, or rationing at the pump. From the vast number of white heads in Washington, it might be assumed that a few would retain memories of the 1930s, when a domestic fuel from the farms of the Midwest was blended into motor "gas" for about 2,000 service stations. Also forgotten, if ever known by the legislators, is the fact

1

that about four million automobiles, trucks, and farm tractors in the United States, Europe, Latin America and the Philippines, have run, with no major mechanical alterations, on fuels derived from non-petroleum sources. This all occurred before 1938.

Peacetime Use of Alcohol

In peacetime, the use of alcohol as a motor fuel was sometimes fostered by its low price in the United States. Abroad, governments encouraged its use through high import duties on oil, subsidies for producers of "power" alcohols, and other supports of alternative fuel supplies, in the likely event of military emergency. In the United States, interest in alcohol fuels was intensified by rising gasoline prices, by stagnation in the liquor industry and by glutted grain markets, when corn sold for a dime a bushel, and farmers burned their grain to heat their houses or trucked it to a distant distillery, to barter it for motor fuel and some fertilizer and cattle feed.

Another strong force for alcohol fuel appeared in the first serious ecology movement to come to the attention of the legislator: professors at the University of Iowa, and other grain state agricultural institutions, initiated legislation conducive to alcohol-blending with gasoline. They did not know that gasoline exhaust, inhaled over a period of time, was hazardous, but they did know that the replacement of the horse by tractors removed an important fertilizer from the farmers' dwindling store. Spent mash from distilleries was to become the manure from the iron horse.

It was unlikely that the professors knew that alcohol, even in a one-to-seven mix with gasoline, would reduce harmful exhaust emissions to a point nearly satisfactory to lawmakers 40 years later. This farm-grown fuel was ethyl alcohol (ethanol, or grain alcohol).

Promising Alternative

Current excitement is on methanol (methyl, or wood alcohol), the most promising alternative to gasoline, and a proven superior fuel in the past.

Raw Materials

The raw material for its manufacture are more than adequate: they include coal, wood, municipal solid waste and surplus crops. The plant capacity to convert these domestic carbons to liquid fuels might be constructed in less time and at lower cost than the development of offshore and remote oil wells and their contingent refinery and transport facilities. Moreover, methanol is the only secondary energy fuel that could come solely from an infinitely renewable primary source, the bio-mass: trees and other solar power collectors.

Doomsday predictions have a way of softening with time, but it is quite clear that we are running out of oil and natural gas in the United States. Anxious readers may be consoled that motor fuels used in 1910, and now more desirable than ever, are being re-examined and found more than adequate. Volkswagen engineers have given prestigious endorsement by designing and testing prototype engines for use with pure methanol fuel. The reports demonstrate performance and economy superior to gasoline, without need for emission control gadgetry. Although there is no mention of the strategic value of homemade fuel in case Europe and the Western hemisphere are cut off from Eastern oil, the dark cloud of this possibility hangs over the horizon.

Other Systems

While methanol, and in certain circumstances, ethanol, are practical potential extenders of the gasoline supply, there are

This huge French gasogen unit was not pretty but was simple and accessible. Cylinder at left produced the gas, short cylinder was the purifier, and horizontal tubes were for cooling. (From New Steam Age, *Vol. I, No. 2, April-June, 1942, p. 11.)*

other strategies and systems. Some are well-known, like the electric vehicle, and others are only dimly remembered from the past. The *portable gas generator,* seen most frequently overseas, was one of the latter. It appeared all over Europe, on taxis, trucks, and private cars, to consume enormous quantities of wood chips, charcoal, coke, lignite, and plain coal. This miniature portable gas works, with a bit of contemporary technology, might be reduced in size and improved in convenience and efficiency to become the omnivorous goat of autodom, in case of an emergency. This unsophisticated creature has had little attention since 1946.

The layman's confusion about the energy problem is excusable. The magnificent, nostalgic steam car was revived in the early 1970s with much fanfare, because it was discovered to be nonpolluting. However, when it proved to be an oil hog, it was re-interred quickly.

Various Problems

Many other unorthodox automotive engines, and a few alternate fuels, such as methane and hydrogen, have been exposed to an eager public, but the engines can't be produced reasonably, and the fuels can't be burned, or shipped, with economy and safety, and without disruption of a huge segment of the personal and commercial lives of wheeling Americans, particularly those in the one-sixth part of United States population connected with the traditional, tightly controlled systems of distribution in automotive, fuel, and associated service industries. One exception — there may be others — is methanol, a synthetic fuel, blended with gasoline.

If a very small proportion of drivers in the United States can be diverted, by this work, from the myth that gasoline is the only fuel that will propel them, the book's aim will be fulfilled.

1

The Drinking Automobile: Another Diet?

The federal government and the oil companies are presenting two choices to American drivers, as the price of gasoline steadily rises. One is to drive less; the other is to pay more.

Both are unsatisfactory to a driving public which has made the family car so much a part of its way of life. This book is written to show that there are other alternatives if, instead of relying on gasoline alone, we look at other fuels. Some of them have proved to be completely satisfactory in the past, and were shelved only because gasoline was so cheap and available.

Methanol

One of them is the synthetic fuel *methanol,* which we can make from renewable and waste materials, and which can be used either straight or in a mixture with gasoline.

A mixture of gasoline and methanol, with as much as 15 percent methanol, can be used without changes in the internal combustion engine in our cars, and use of this mixture will reduce our consumption of and dependence on gasoline by that percentage.

With engines designed for its use, straight methanol can be used as fuel. It works better than gasoline — witness the 200 mph cars fueled principally with methanol and racing at Indianapolis and other major tracks. It permits smaller, lighter engines.

The gradual substitution of this fuel, once called wood alcohol or methyl spirits, for a part of the gasoline now imported, is an answer to our dependence on foreign oil supplies. It is one of the few fuels for the traditional internal combustion automobile engine that can be made from coal without a threat of air pollution. Better yet, it can be made from rubbish, solving a problem for municipal treasuries depleted by rubbish disposal headaches. Or it can be made of wood wastes normally left in the woods or burned as a fire-protection measure.

Why is knowledge of this excellent alternative fuel limited to racing drivers, the insiders of the automotive manufacturers and petroleum chemists and lobbyists? It is convenient to answer that there has been no advertising, and that the fuel is in relatively short supply. Other reasons will be given later, but the first must be considered valid: The production of methanol in the United States is only one billion gallons per year — 1 percent of gasoline consumption.

Gradual Switch to Methanol

This should not block a gradual switch to use of methanol in our cars. An industry that can produce a billion gallons in a year can raise its output at a rate of 5 to 10 percent a year, when there is the incentive of a greater market created by the rising price of gasoline, and there is an abundant supply of low-cost raw materials.

Europeans born before 1940 know firsthand of the value of gasoline substitutes such as alcohol fuels and the gas produced by a wood-gas generator for vehicles, but in the United States the lack of knowledge of alternatives to gasoline is almost as complete as the Egyptian slaves' inability to read and write. For a country whose wealth is largely based on technology, this small but costly bit of illiteracy, in the zone of our most intimately loved vehicle, is surprising.

Even more surprising is the fact that the various federal energy bills under consideration provide only a small, undefined amount of support for the development of liquid fuels from coals. The 1973 Nixon energy program emphasized natural gas well development and elimination of restraints on producers, oil drilling on the continental shelf, geothermal leases on government lands, revival of the coal industry, continuation of a 20 percent oil depletion allowance, encouragement of nuclear generating expansion and conservation by voluntary sacrifices. The Ford administration proposals are similar.

No Long-Range Answers

It is doubtful that either can be considered a program offering long-range answers. The considerable difficulties of offshore drilling, both physical and political, and the announcements of some geologists that early estimates of American gas and oil reserves were far too rosy, have

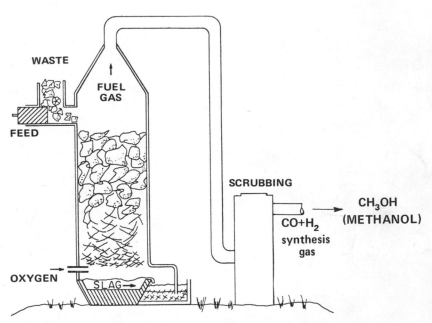

An oxygen waste disposal furnace, showing the basic method of gas production.

dampened our hopes for the ever-flowing pipeline. We need the 29 percent of our oil imported (1973) from other countries, and we will need more as time goes on, unless our dependence on it is changed through bold, innovative thinking.

Such thinking has been lacking in this country in the development of an alternative fuel. Technologists know that petroleum contains many of the raw materials for the plastics, synthetic fertilizers and chemical products essential to our economic health, and most of them agree that petroleum is too good a material to be burned. Despite this, they have been reluctant to develop a fuel manufactured from renewable resources, for fear that it could not compete on a cost basis with fossil fuels. The era in which a competitive fuel cannot be developed, produced and sold is at an end.

The Internal Combustion Engine

This is a good time to examine one of our largest guzzlers of fossil fuels, our inefficient but beloved friend, the internal combustion engine.

This engine is a remarkably versatile performer, exerting itself imperceptibly to leap from sluggish urban traffic to mountain climbing at 60 miles an hour. Its tastes for various fuels are unbiased. It runs fairly well on natural gas, coal gas or sewage gas, on its weekday diet of gasoline or diesel oil, and, on a spree at the race track, drinks high-proof vodka or methanol (ethyl and methyl alcohol).

Such flexibility is surprising, in view of the development of the engine as a constant-speed power source to run such machines as saws, forge blowers, butter churns and hammer mills, and to be fed solely on illuminating gas (made from coal) through a pipe.

This led to one of the first substitutions in fueling methods. The piped-in gas restricted Lenoir's first 100 internal combustion engines to stationary uses. But when one

A factory-built gasogen on a French sedan, installed in the trunk space. Upper curved part stores charcoal, roof rack stores more. On front fenders are the gas cleaners and coolers. This is a typical World War II installation.

was made portable for use in a cart, it was necessary to pump coal gas into a pressure tank on the cart. While this was an inconvenience, it was a minor one when compared with others, such as pushing the cart to start it, the roasting-hot engine under the floor and the lack of brakes.

Gasoline: The First Substitute Fuel

Gasoline then became the first substitute fuel, although it was considered extremely hazardous to store and convey, and it was atrociously reluctant to ignite in the cylinder. Gasoline, by circumventing the problem of compressing coal gas into a pressure vessel, brought the horseless carriage a little nearer to the fringe of public acceptance.

Civilian adoption of the internal combustion engine as a replacement for the horse was slow. The engine much more quickly was seen as the best method for powering such instruments of war as the truck, the tank and eventually the airplane. Thus, as gasoline replaced oats as the key to military mobility, the nations of Europe learned that control of petroleum supplies was essential to military success.

Between world wars, many countries not endowed with oil resources required that alcohol be added to gasoline to extend their expensive imports, and they also taxed fuel and cars to reduce the flow of foreign exchange.

Charcoal and Wood Waste

In another development, the portable, vehicle-mounted gas generator was refined, and this permitted substitution of charcoal or wood waste for 95 percent of the users' gasoline supply. Both alcohol blending and gas generator development and promotion were military endeavors, carried out by government decree on public vehicles in the 1930s.

Such developments hardly touched this country. We had a domestic supply of oil that was, until a decade ago, considered adequate for any emergency, so we gave hardly a thought to alternative fuels. Only an agrarian movement in the 1930s, in the Midwest, using surplus grain, ventured into this field. Alcohol was used in blends by about 2,000 filling stations for a few years.

The idea didn't spread. The popular attitude at that time was expressed in an American Petroleum Institute publication titled, "Power Alcohol, History and Analysis." That publication stated:

> "Pretentions that alcohol is needed as a substitute for irreplaceable oil supplies are answered by the fact that petroleum reserves are greater today than ever before, while conservation methods are improving rapidly. Also that methods already developed of synthesizing oil from coal assure a continuing supply of oil as far into the future as any need can be foreseen, at cheaper prices than are in prospect for alcohol."

A Different Forecast

Only a few years later, at a symposium of the institute, members heard a far different forecast, but one that received little attention. That forecast was that there would be a depletion of oil supplies in the late 1960s, followed by the raising of "grave policy questions with regard to the future of the petroleum industry." The speaker was a geologist, Dr. M. King Hubbert, then in the employ of Shell Oil Co. When the proceedings of the symposium were published, Shell elected to omit this warning from the paper.

Dr. Hubbert's prediction missed by only a few months. Production of U. S. oil, as recorded in the "Oil and Gas Journal," hit its peak in January 1970 and has dropped consistently ever since. Dr. Hubbert, now lecturing to more receptive ears than 20 years ago, says, "In about 30 years, American oil will cease to be a major energy source, and in 50 years the world's supply will have run its course."

If Oil Shipments Cease?

The nation's oil problems might be intensified long before then. With tensions in the Near East and Middle East remaining high, it is prudent to assume that oil shipments to the United States, and perhaps the West as a whole, would cease with the outbreak of a war, regardless of the alignment of powers. What steps can be taken by government and private enterprise now to forestall a paralysis of part of our private transportation equipment?

One obvious step is to reduce our gasoline consumption, and means of doing this have been put forth frequently in recent years. The recommendations of President Ford's "Project Independence 1980" include raising miles-per-gallon to an average of 19.6. This could be done by reducing curb weight of vehicles, engine size and accessory loads, by using radial tires to lower tractive effort, by installing positive transmissions, and by improved streamlining of body contours.

But can we expect a reduction in automotive fuel consumption? The answer is no. Although there was a temporary slowdown of petroleum consumption in early 1975 due to the nationwide depression, economic recovery is expected to bring a constant increase in the number of vehicles, a resurgence of highway building that will extend the average length of trips, and an increase in gasoline consumption that will erase the reductions of voluntary or price-induced curtailments of driving.

In Event of Emergency

Our own oil may run out well before the magic year 2,000. But what if an emergency arises before 1980?

Methanol could avert the enormous social and economic upheavals that would result in this country. If blended with 85 percent regular gasoline, it can be used in today's automobiles without modifications. It is not difficult to make from wastes, coal, wood, or surplus bio-mass of any kind.

But this nation cannot convert to its use without planning such a move now. Plants to make methanol do not rise overnight. Even if a decision to go ahead with a nationwide methanol project were made tomorrow, years would be needed to raise the output to useful levels, particularly if present fuel producers were left to make the decisions.

A Long-Range Decision

Although rationing and the imposition of tariffs and import taxes on oil, and graduated taxes on high-horsepower vehicles are other obvious immediate steps toward conservation, the only rational long-range solution to the problem offered to date is the expansion of methanol production. This is possible through state legislative action and consumer pressures on educational and research agencies at the federal level.

Another, longer-range activity in fuel conservation deserving more public encouragement is the development of diversity in vehicle design, particularly focused on the electric car as a second family car. The wood-burning gas generator (gasogen) is another more limited prospect in farm and forest planning.

These methods, although unsatisfactory in one way or another to a nation coddled in luxury motoring habits, are tested and workable. The electric car has a long history in this country. The gasogen, well designed and fed by government-aided charcoal distribution, gave Brazil its lifeline transportation during World War II. An effective Brazilian General Motors organization soon adjusted to the idea that it was better to be in the gasogen business than in no business at all.

Such alternatives should be studied, since if oil imports should dry up, for any reason, no single substitute fuel, no one alternative engine, would solve our problems. A diversity of fuels should be planned, each suited to the geography and economy of areas. The single-fuel economy to which we are addicted is neither necessary nor healthy, but it serves perfectly the conservative monopolistic instincts of two huge interdependent industries.

Automotive Progress

There are many flat periods in the history of technology, such as the plateau of 100 years between the atmospheric engines and Watt's first true steam engine. Automotive progress, examined critically, enjoyed a similar hibernation of about 60 years, between the self-starter and the rotary engine. There were refinements — four-wheel brakes, automatic transmissions, etc. — but they were generally imported from Europe after a decade of trial, negotiating, or treading water, until coercive pressure of some kind forced adoption in Detroit.

The fuel shortage stimulated some notable progress. The stratified charge, the effect of layers of rich and lean gas in the cylinder of the internal combustion engine, was described by Nickolaus Otto as a means of absorbing the shock of explosion on his 1876 "silent" engine, but only a few inventors, usually on their own hunches, examined the shapes of combustion chambers and the phenomena of flame behavior and exhaust emissions during the next 90 years. Then the Japanese Honda engineers turned their labors and

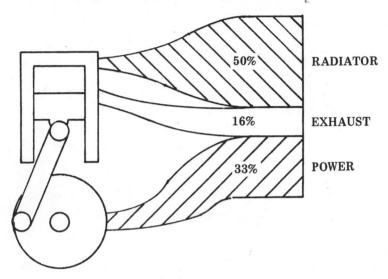

Diagrams illustrate what has happened to the power of automobile engine with the introduction of emission kits to reduce air pollution. Top diagram shows that 33 percent of total heat content of fuel is used for power, but that, when emission kit is added, in lower diagram, only 24 percent of power is used productively.

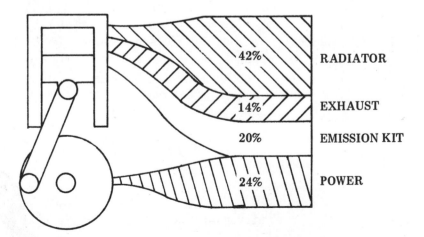

keen intuitions to this area — and out came the *Civic*, the sporty little engine with controlled carburetion and combustion that needs no emission-control diapers.

What Can Be Done?

There are similar steps that could be taken to increase the efficiency of American cars.

Reduce weight. Unnecessary weight in a car lowers performance. While this is known by all automotive engineers, many of the cars of the early twentieth century were lighter in pounds per horsepower, and in the proportions of light materials to steel, than present cars. The cost, in energy, of recycling junked and rusted steel shells is a hidden horror of programmed obsolescence originating in the styling parlors of Detroit.

Improve Styling. Sometimes styling is said to reduce air resistance, thereby improving mileage. Has it?

The Rumpler sedan of 1921 had an average drag coefficient of 0.5, meaning half the resistance of a block of the car's dimensions. To achieve reasonable mileage at high speeds, a coefficient of 0.25 to 0.30 is urged. Certain U. S. dragsters and European sports cars have gone down to less than 0.2, and the standard French Citroen sedan has had 0.3 for years. American cars have shown little progress from the Rumpler sedan of more than 50 years ago. Present body styles of the American cars have an average drag coefficient of 0.47. The result of our neglect of genuine streamlining may be a few hundred gallons of gas per year per car extra, and extra billions of gallons wasted by the nation.

Innovative Contributions Needed

Such short-term conservation projects, with enforced speed limits and the rising price of gasoline, could contribute to the reduction of consumption. But they will be inadequate to meet present growth or future emergencies.

More innovative steps are needed. In the early 1970's, the technical literature began to show signs of interest in using coal and synthetic fuels for transportation. While the automotive and petroleum industries would be expected to take leadership in this field, few projects are sponsored by them. Instead their engineering members seem to be occupied with the specifics of fending off irritated consumer groups, meeting environmental standards and listening for the diminishing clinks of management's cash registers.

Several alternatives to the gasoline economy have been proposed. Chief of these is hydrogen, an odorless, explosive gas having half the energy content of gasoline. It would require a redesigned engine and pressurized fuel system, a totally separate fuel dealer distribution system, and to provide our fuel supply, a nuclear-electric generation plant to convert water into hydrogen and oxygen — economically.

Another alternative is methane, which, while it is feasible as a stationary engine fuel, presents the same handling problems as hydrogen, with the added deterrent of being slow to produce by usual biological means.

On-board gas generators, although bulky in present forms, offer a practical homemade substitute for liquid fossil fuels. Problems in their use include finding supplies of adequate, uniformly sized fuel, and providing constant maintenance. This substitute allows alternate gasoline or wood gas operation, with about 50 percent efficiency with the latter fuel. Hydrogen and methane require continuous use of the single fuel due to engine modifications required.

2

Methanol: Instrument of United States Oil Conservation

If critics of modern technology distrust the speed of its progress, they should find consolation in the fact that it often goes in circles. A century ago, almost all United States steam transportation was powered by a renewable fuel — wood. Since 1970 there has been talk, research, and action that will inevitably lead to partial return to wood as the raw material of fuel for the automobile. Municipal rubbish, which is very difficult to give away, is rich enough in paper (wood pulp) to be converted into methanol, a motor fuel of first quality.

What is Methanol?

Methanol is the international chemical name for wood alcohol, or methylated spirits. It is a widely used solvent and raw material in the chemical and plastics industries. It is the low-priced antifreeze, poisonous, like gasoline, but more often hazardous because it may be confused with ethanol, or "grain" alcohol, and drunk. It is practically odorless. It may be made from coal, wood, waste, or any material containing carbon, but, like many other commodities, it is presently made from the most economical source, natural gas. It should be handled like gasoline, although it is somewhat less hazardous. In an engine it burns cleanly, without depositing carbon. It is the only fuel that a wise yachtsman will use in

18

his galley range, as its exhaust is only water vapor and carbon dioxide, identical with the yachtsman's own exhalation. If there is a small fire, a pan of water will extinguish it, not spread it, as occurs with a kerosene or gasoline fire.

Methanol-Gasoline Blends

Up to 15 percent methanol can be added to gasoline in current cars, without adjustment of the engine, and with noticeable improvement in exhaust quality, economy and performance. Methanol has an octane rating of 106, compared to typical gasolines of 85 to 100. It prevents knocking, or "pinging," common with unleaded fuels, and it alleviates "running on," or "dieseling," when the ignition is switched off. These practical and well-documented qualities, added to the virtue of reducing national fuel dependence on the Organization of Petroleum Exports Countries, have made methanol the leading potential candidate for the motor fuel of the immediate, as well as foreseeable future.

It would be wise to consider both economics and available technology before launching any substitute fuel in a market now calling for 100 billion gallons of gasoline (1972). Methanol production for the same year was only 1 percent of this, one billion gallons. The Atomic Energy Commission, in a report titled "Production of Methanol from Coal for the Automotive Market" in early 1974, suggested a demonstration plant to produce 5,000 tons per day of methanol, capable of raising present production capacity by 50 percent.

No Action Yet

Despite the national commitment to develop coal and clean energy fuels, the AEC report has thus far produced no action. The report included a computed cost, before mark-up and taxes, of 13.5¢ per gallon for methanol made from coal. This is very close to the historical price of gasoline as it leaves the refinery and is about 6 cents less than the usual methanol price. It is possible that the report constituted a battle flag to the watchful and powerful forces of the petroleum industry, which has everything to gain by preserving the status quo. Until 1973, methanol's potentials were known only to an inner circle of chemists and automotive professionals, who can be very tight-lipped at times.

A Different Gas Was Tried

The energy crisis stimulated many groups of ecologists, youthful nonconformists, and nature buffs to study methane, a gas arising from decaying organic matter of all kinds, as an all-around substitute for petroleum fuels. While diligent craftsmen could retrieve enough methane from septic tanks and manure digesters to cook a few meals or run a little gas engine, not many attempted the awkward conversion of cars to this fuel.

The popular revival of the idea of a practical motor fuel from renewable, organic, non-oil sources came with a story on the front page of the *New York Times,* Dec. 31, 1973. This article was abstracted from the December 1973 issue of *Science,* in which T. B. Reed and R. M. Lerner of the Massachusetts Institute of Technology reported on their experiments with methanol as an extender of gasoline supplies in their cars and those of their colleagues.

Two Ways to Use Methanol

Methanol is used as a motor fuel in two ways, each having distinct advantages from the physical standpoint, and some minor disadvantages to the consumer. We shall consider first its use in a "blend" of 5 to 30 percent methanol to 95 to 70 percent gasoline, which is the range chosen by Reed, Lerner, and friends for testing.

Nine cars, vintages 1966 to 1972, with horsepowers from 57 to 335, were tested on blends of 5 percent to 30

Dr. Thomas B. Reed's Toyota, which, when this photo was taken, had run 30,000 miles on methanol blends without a hitch.

percent methanol. The cars were unmodified and tests were over a fixed course under standardized conditions. A summary* of findings was that (1) fuel economy increased by 5 to 13 percent; (2) carbon monoxide emissions decreased by 14 to 72 percent; (3) exhaust temperatures decreased by 1 to 9 percent; (4) acceleration increased up to 7 percent. The elimination of knock and of "dieseling" was noted, even on the lowest 5 percent methanol blend tested. The latter improvements were unexpected, but were explained tentatively by the possible dissociation of methanol in the car's cylinder, with attendant absorption of heat energy, quenching early combustion. Simply stated, it burns "cool."

There are problems in the storage and dispensing of mixtures of methanol and gasoline. Gasoline containing 10 percent methanol will absorb 0.1 percent water — ten times as much as gasoline alone. Thus, in a system using the blend

---

*"Improved performance of Internal Combustion Engines Using 5-20% Methanol." R. M. Lerner et al., Lincoln Laboratory, Massachusetts Institute of Technology.

continuously, normal amounts of water formed by condensation are carried away to the engine — the "dri-gas" effect.

However, in wholesale storage and distribution of gasoline, water is sometimes used to displace gasoline to prevent the possibility of vapor accumulation and explosion. Residues of this water, with normal leakage and condensation in tanks, are easily separated by traps in transferring gasoline. But when 10 percent methanol is present, the water will desorb the methanol in large amounts.

At freezing temperatures (0°C), less than 10 percent methanol is soluble in some gasolines. However, it would appear that changes in the handling of bulk fuel or in the point at which blending occurs would solve the water problems.

The separation of the components may be prevented by the presence of small amounts of higher alcohols in methanol fuels. Curiously, it is easier and thus less expensive to produce methanol with these other alcohols — ethanol, propanol and isobutanol — in it, and the output of a manufacturing plant is increased by 50 percent.

Difficulty in Cold

Reed and Lerner warn experimenters in northern climates not to use blends of *pure* methanol with the temperature much below 30 degrees F., unless they are prepared to *stir* the tank frequently. The penalty: The methanol-water mixture settles to the bottom of the tank, and when one attempts to start the car, ignition of *cold* methanol will fail.

Experience has shown this is no major problem. A reassuring report resulted from extended testing of blends in England in the 1930's. There, at the Billingham Works of Imperial Chemical Industries, methanol-gasoline blends were dispensed from the works' garage pumps for over a year and a half, including a cold winter. The supposed "hydroscopicity of alcohol fuels and the resulting dangers of separation" did not appear, they reported.

"It is true that, under laboratory conditions, alcohol-petrol blends absorb water vapor from the air when freely exposed to the latter and eventually separate into two layers, but, in actual practice, motor fuels are not freely exposed to the atmosphere, and, as a result, can often be used with satisfaction."*

Add Methanol Gradually

A second warning when using a blend for the first time: Add enough methanol to the tank, when taking a trip, to make a 5 percent blend. This will clean out any water in the fuel system. After ten miles, if there is no sputtering, add enough more to make a 10 percent blend.

Critics of methanol blends, often retired engineers writing letters to the newspapers, enjoy pointing out that the alcohols have lower heat contents than gasoline, and therefore *must* reduce the power and mileage performance. They are correct in the first premise; but because alcohols induce better combustion of gasoline and lower the temperature in the cylinders, the blends, up to about 20 percent for methanol, actually improve power and economy. This kind of synergism prevails frequently in the realm of biology, but is seldom perceived or comprehended in mechanics.

Methanol is not highly toxic, but 30 to 100 cc. can be lethal if ingested. It is less dangerous than gasoline if inhaled, and far less toxic than the two popular household cleaning fluids, trichloroethylene and carbon tetrachloride. If it came into general use, its chief hazards would be controlled by label warnings (not mentioning the word alcohol) and *not* siphoning fuel by mouth, as is sometimes done in emergencies with gasoline.

Another unsuspected hazard is that of carrying a leaking can in an automobile. Being odorless, the leaking vapor might not be detected in time to avoid considerable inhalation and to avert tragedy.

The Principles of Motor Fuel Preparation and Application, by Alfred W. Nash and Donald A. Howes. See bibliography.

A Critic Answered

The Reed and Lerner report evoked some studiously negative predictions from researchers and officials connected with the petroleum industry. First came a critique by Dr. E. E. Wigg of Exxon Research and Engineering Co. in which he posed seven points of disadvantage for methanol. His report included most of the objections ever voiced about methanol, and therefore it is laid out below in some detail with comments by this author.

1. Methanol gave good test results in older, less efficient cars. The MIT testers should have chosen a higher proportion of late-model cars (then, 1972), on which the gains due to methanol were claimed by Wiggs to be insignificant.

   Reply: Later tests by the MIT group, as well as by Volkswagen researchers, show similar excellent results on 1974 and 1975 cars. T. B. Reed reports (July, 1975) 25,000 trouble-free miles on his 1974 Pinto and 35,000 miles on his 1969 Toyota, on a 10 percent methanol blend.

2. Three highly technical paragraphs showed the difference between octane blending values (OBV's), research octane number (RON), and motor octane number (MON), ending with the conclusion, apparently derived by interpolation of other researchers' papers, that a fuel saving of less than 2 percent would accrue to a car when the octane number is increased by one.

   Reply: The inference is that the mileage figures given by the Massachusetts testers were flukey. Even in the highly unlikely case that errors were made, the merits of methanol blending speak out in an elementary consumer test. A 1974 car is found to knock badly on acceleration with the specified unleaded gasoline. However, when a gallon of methanol is added (to 9 gallons in the tank) the knocking disappears. That means increased power, which usually translates into increased mileage. Whether it is 2 percent or 5 percent is not so significant as the 10 percent reduction in petroleum consumption.

The Volkswagen researchers report that both RON and MON are unreliable in comparing methanol to gasoline, since faster burning produces a pressure rise which could be confused with knock, but is not knock.

3. Methanol and gasoline stratify in tanks. (Wiggs calls this "phase separation.") This problem, mentioned earlier, is given great importance. Wigg reports his laboratory test showing that less than .1 percent water in a blend causes separation. (This is a standard 15 percent blend.)

Reply: .1 percent means 2.5 ounces in a 20 gallon tank. Any car using gasoline alone can be immobilized by *one* ounce of water frozen in the gas line; the problem is nasty, but not insurmountable.

4. There is an increased possibility of vapor-lock. This unpleasant accident occurs in very warm weather with inadequate cooling of a vehicle's fuel pipe by the air stream. Placed under reduced atmospheric pressure by the suction of the fuel pump, the fuel vaporizes. This slug of vapor in the line blocks the liquid fuel, and the driver is forced to wait until natural cooling and condensation occur, or to look for a pan of cold water. Wigg computed various vapor pressures of blends, and offers a number of impractical remedies for reducing the vapor pressure by altering the components of the gasoline fraction. No field tests concerning vapor lock are reported.

Reply: Gasoline refineries already change the characteristic spectrum of volatile components in their products with the season, to reduce the incidence of vapor lock in summer and to facilitate starting in winter. Methanol users have reported no problems of this nature. The problem, for either gasoline or blend users, is one that the manufacturers know how to solve, by fuel pump or piping relocation. They refrain from action for the sake of economy. The MIT group, using methanol blends, reported no cases of vapor lock in two seasons of Boston climate, and Dr. Reed, on a vacation tour into southern states in July, 1975, had no difficulties of this nature with his Pinto.

5. Methanol reduces fuel economy. Exxon tested three vehicles, running on two fuels: unleaded gasoline and a 15 percent methanol blend. Although arrived at through complicated manipulation of the "equivalence ratio" of fuels, their conclusions were simple: By the time that methanol is available in adequate quantities, all cars will be so efficient on gasoline alone that methanol will not improve their economy.

Reply: No comment.

6. Methanol increases exhaust emissions. Tests indicated that emissions on newer cars were not significantly reduced, and that additional potential pollutants, called *aldehydes,* were created by methanol blends. Recommended: Further investigation.

Reply: Aldehyde emissions are only one tenth of hydrocarbon emissions and they degrade much more rapidly in the atmosphere. Volkswagen investigators, in both field and laboratory tests, found that blends lowered emissions significantly. With 100 percent methanol, they report that graduated additions of water brought reductions in nitrogen oxides, and up to 40 percent less aldehydes.

7. Methanol reduces car performance. Testing "prior to warmed up operation in the 1975 federal test procedure" showed that cars on 15 percent methanol blends stalled, hesitated, and backfired more than those on pure gasoline (except for a 1973 car, which stalled on both fuels equally).

Reply: The number of occurrences per test was statistically too low (1 to 4) to be conclusive. The faults are typical of cold engines, lean fuel mixtures and impatient operators, not necessarily of methanol blends.

Dr. Thomas B. Reed of MIT dispensing methanol for blending.

Writing in *Science* magazine, a critic of Wigg's critique pointed out what must be obvious here. Wigg ignored the fact that 15 percent methanol by volume added to gasoline would decrease by that amount our dependence on gasoline. He also ignored the fact that adoption of blends might reduce or eliminate the need for catalytic converters on new cars. While the first omission is an incontrovertible oversight, the second was challenged by Wigg, on the strength of the stringency of future standards for emissions. Since the entire automotive industry resists the early enactment of strict clean air legislation as cats resist the warden's net, and the catalytic converter is being scrutinized as a producer of acid fumes, this question might be held for further investigation.

Corrosion Problems

One possible complaint against methanol as a blender that was aired in the 1930's but has remained under cover thus far was that methanol was corrosive to certain materials in a car's fuel system. At the time of trial, carburetor floats of cork, and gaskets sealed with shellac were easy game for alcohol. Present metal floats and synthetic cements resist the solvent action of alcohol. Carburetor parts are made of zinc die castings, sometimes aluminum. The impurities in both metals in earlier days were conducive to "intergranular crystallization" as a result of aging. This crumbing destruction could be accelerated by the presence of alcohol and water, but the problem no longer exists. Lead, tin and magnesium are attacked by methanol, but there should be no opportunity of exposure in the combustion zones of an engine to these metals. Iron and steel are quite immune, as are brass and bronze.

100 Percent Methanol Fuel

North American and European ventures into alcohol fuels, both in recent times and in the 1930's, depended on blends ranging from 2 percent to 95 percent alcohol. The motivations were numerous, but the underlying reason for adhering to blends was that no alteration of the automobile's engine was required. However, about 1910, pure alcohol was an established alternative fuel for the horseless carriage. Assuming the availability of methanol at the service station around the corner, what kind of an engine would be designed for its maximum benefits?

First, the amount of air consumed by burning methanol is reduced. While the ratio for gasoline is 14 weight units of gasoline to 1 of air, for pure methanol it becomes six parts fuel to one of air. Carburetor jets must be changed.

Second, heat is needed at the methanol intake to

vaporize it. A loop of exhaust pipe around the air intake will do the job.

Third, an initial cold start requires an electric glow-plug or a highly volatile primer fluid, such as gasoline or ether. Priming fuel is fed to an auxiliary valve from a small tank. Such a conversion might cost around $100, if made on an existing car. If incorporated in a production model, the cost would be one-tenth, or nothing, as the trade-off could be the elimination of emissions control devices — all of them.

Santa Clara, Calif., has a city-owned vehicle on test with such a conversion. Test data showed emissions, compared with gasoline, of one-twentieth of the hydrocarbons, one-tenth the carbon monoxide, and about the same oxides of nitrogen. A 1972 Gremlin made the emissions recorded in the accompanying chart (grams per mile) on gasoline and methanol fuel. The 1976 federal standards are placed at bottom line for comparison.

	Emissions, Grams/Mile		
Fuel	*HC, unburned*	*CO*	*NO_x*
Gasoline	2.20	32.5	3.2
Methanol	0.32	3.9	0.35
Federal standards	0.41	3.4	0.40

Methanol burns without misfiring at leaner mixture ratios than gasoline. This does not mean that it uses less fuel — in fact, methanol quantity consumed will double — but it indicates more complete combustion and less pollutant exhaust than gasoline. Temperature of exhaust is 100 degrees Centigrade less than with gasoline, and spark timing may be later (more efficient) with methanol, because of its higher flame speed. Higher compression ratios are feasible without toxic additives, and "dieseling," or running on after the ignition switch is turned off, is eliminated because of the higher heat of methanol evaporation. (Methanol is not a good diesel fuel.)

For the ordinary driver, these qualities will be apparent in improved acceleration, a lighter car and engine, fewer parts to service (no catalytic converter, etc.), but a car that needs about twice the fuel-tank size for the same mileage as gasoline. This problem caused the sad failure of the Novis, an advanced design of Indianapolis "500" race entries, designed to run on 100 percent methanol, but unable to finish the race at the speeds of their qualifying runs. The rules limited the fuel used — in gallons instead of in heat units.

Volkswagen Tests

Volkswagen's Research and Development Division gained test experience on two new model VW's and on a Daimler-Benz 4.5 liter V8 engine during 1974. Tests compared all pertinent data for gasoline and pure methanol. Nine months of VW durability testing, with 30,000 km accumulated, show no serious defects due to methanol use. A paper on this program was presented* at the Second Symposium on Low Pollution Power Systems (NATO/CCMS) on Nov. 4, 1974 at Dusseldorf, Germany.

Conclusions reached in the paper were positive, and, for a scientific team, enthusiastic. "Of the possible alternative automobile fuels, methanol takes a leading place because it can, on the one hand, be produced on a non-mineral oil basis, e.g., from coal, at a comparative price and can be stored and distributed as a liquid fuel in the same way as gasoline." In addition, improved emission and performance can be obtained with methanol.

Methanol Production

The Volkswagen researchers' statement on comparative price with gasoline is important because a detailed study of production methods in Germany and the United States was

---

*"Comparative Results on Methanol and Gasoline Fueled Passenger Cars," by Herbert Heitland, Winfried Bernhardt and Wenpo Lee. R & D Div., Research Dep't., Volkswagenwerk AG, Wolfsburg, Germany, 1974.

VW car used in 1974 tests by VW's Research and Development Division.

made, and costs were predicated on conventional technology, as well as on nuclear energy for process heat. Because of traditionally higher prices for motor fuel in Europe, the advantage of methanol was expected, but in the United States, petroleum economists have assumed, until the present time, that alcohol could not compete with low-cost gasoline in the free market. Costs of methanol, as computed for Germany by VW researchers, varied widely according to the method of production used. Prices in the table below are at refinery before tax, except x and y, which are pump prices.

Cost of Fuel, $ per MMBTU

Fuel	In Germany	In U. S.
Methanol	2.50 — 5.75	1.87 — 3.06
Gasoline	4.00 — 5.5*	5.24 (x) — 6.11 (y)

*(March, 1974) (MMBTU = 1 million British Thermal Units)
　x = gasoline @ $.60/gallon, July, 1974.
　y = gasoline @ $.70/gallon, forecast for late 1975.

It is clear that in Germany methanol can approach a competitive price with gasoline, and that in the United States there is an equally promising situation.

Methanol Production Chemistry

The first step in methanol production is making "synthesis" gas. This consists of carbon monoxide and hydrogen, the latter being provided by the admission of steam over the combustion zone.

$$C + H_2O = CO + H_2$$

This same formula will appear again (chapter 3) as the product of portable gasogens. It is, perhaps, the oldest industrial gaseous fuel, similar to blast furnace gas, "producer gas," illuminating gas, water gas, etc., and the "city" gas supplied to customers in the nineteenth and early twentieth centuries.

Natural Gas Used

The source of methanol is now natural gas, oxidized in part by steam. Of course, our dwindling natural gas supplies will be diverted to essential or more politically expedient uses and may not be available for expanded methanol production in the future. Near Eastern oil wells presently "flare off" — that is, burn the natural gas at the wellhead. If this huge amount of gas, now wasted, were liquified to methanol, it could be shipped in ordinary tankers to customers on this continent, at an estimated cost of less than 10 cents per gallon or $4.20 per barrel.

The manufacture of methanol has evolved in the following stages: the primitive charring of wood, in making charcoal; then, in the twentieth century, distillation in a metal retort and the condensation of acetic acid, tars, and the "methylated spirits" or wood alcohol. The synthesizing process was discovered about 1925. The relatively simple reaction, starting from the producer gas above, is:

$$CO + 2\ H_2 = CH_3OH\ \text{(methanol)}$$

It was accomplished only at extremely high temperatures and pressures; 4400 pounds per square inch (psi), and 200°C (392°F). A recent (1968) process developed by Imperial Chemical Industries allows a lower pressure (730 psi) and temperature of 250°C (482°F) with a copper-based catalyst.

Wood was the principal fuel of the United States before the time of Col. Drake's probing of the earth's substrata for oil in 1859. Now it has diminished to a fraction of a percent. However, one of the progressive forest products companies of the Pacific Northwest has predicted, with considerable research data to back its statements, increased rather than diminished yields from commercial forests, which cover 23 percent of the United States. Although waste wood has declined by about 70 percent in 30 years, there is still available about 3,800 cubic feet of waste per acre in wood-cutting operations in western Oregon. Reed and Lerner estimate that if all our waste biological growth (500 million x 10^6 dry tons/year) could be converted to methanol, it would supply 10 percent of present total energy consumption in the United States. (Transportation uses 25 percent of total energy.)

Current Projects

Maine has 5.5 million acres of forests afflicted by spruce budworm that is killing the trees, and a methanol plant is being considered for converting this carbon supply to liquid fuel.

Morbark machine harvesting spruce in Maine experimental cleanup operation.

In addition, all of Maine's forests could yield wastes that would supply up to a billion gallons of methanol each year, an amount matching the present total United States annual production. Although it is not as efficient to produce methanol from wood as from coal — wood-produced fuel might cost around 25 cents per gallon, compared to about 14 cents from coal — the combined salvage nature of the project and the relative isolation of Maine from distribution centers of gasoline are incentives for local methanol plants.

An even better incentive appears when combustible municipal wastes, that normally cost $3 to $9 per ton for disposal, are delivered to methanol conversion. Americans use about 575 pounds of paper per capita each year, and this could, with other solid wastes, supply about 8 percent of the transportation fuels we need, if converted to methanol. The chief deterrent to wood fuel utilization is that the world's forests are at the greatest possible distances from the centers of population (and utilization), but urban methanol plants place the material source, refuse, and the fuel market in immediate contact.

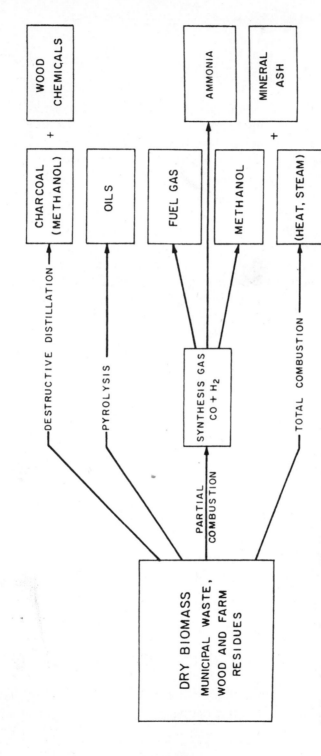

Methanol is one of many products obtainable from dry biomass.

Pilot Plant Considered

Seattle city officials are pushing for a pilot plant to convert solid wastes to methanol, or ammonia to be used for fertilizer. Motor fuel and fertilizer are both in short supply, and the days of landfilling with rubbish are numbered.

In South Charleston, W. Va., Union Carbon and Carbide Co., has a synthesis gas plant operating, as a demonstration, using waste. Pure oxygen is used as the oxidant, rather than air. Deducting the former disposal cost of $10 per ton of rubbish from the cost of methanol, Reed has arrived at an estimated cost of 14 cents per gallon for methanol.

California, where the personal health and economic welfare of the state are closely tied to the amenities of the automobile, has made a serious entry into methanol's viability as an alternative fuel. San Diego's coastal zone commission, with a non-profit group headed by Scott Carpenter, former astronaut, and a Los Angeles assembly-man, William Greene, are using a $1.1 million feasibility study appropriation by the legislature to plan for municipal waste conversion to methanol and its use in state vehicles to lower air pollution and conserve fuel. A San Diego Post Office delivery truck has logged over 24,000 miles on methanol blends, with reported gains of 25 percent in mileage.

Mobil Oil Corporation has been awarded a research contract of $862,000 by the Office of Coal Research to study the manufacture of methanol fuel from coal by the catalytic process. Since the technologies have been known for several years, the project might be compared to a grant to a whiskey distiller for the purpose of developing a soft drink without any habit-forming potential.

Prospects for Methanol Development

One of the global energy-producing companies proudly proclaims, in its TV advertising, that it plans to spend $16

Union Carbon and Carbide Co. synthesis gas plant in South Charleston, W. Va. Uses rubbish for fuel, and oxygen for combustion (Purox process).

billion in developing new petroleum supplies in the next five years. What comparable capital sources can methanol muster?

The abatement of federal and state taxes on methanol fuels would certainly encourage investments in methanol. Educational campaigns through the Environmental Protection Agency, the Department of the Interior, and state conservation groups could be effective. However, the gravest impediment to the free development of methanol as an alternative fuel is that it would be almost impossible to gain a proprietary hold over its raw material sources. Carbon is everywhere, and a monopoly on rubbish, wood chips, moldy corn crops, or marginal coal fields, is not as attractive as control over a nation's oil industry, from well to gas tank.

However, as petroleum prices rise, there will surely come the occasion to re-examine the economics of methanol, according to the formula suggested in the classic text on fuels by Nash and Howes (see bibliography).

Comparison of Costs

Before the discovery that automobile air pollution had an intimate and enormously expensive relationship with public health, the price structure of a fuel system was the sole criterion of its validity. A pre-smog era, mathematical approach to methanol's chances of success in the old, unrestricted market would begin with an analysis of its heat content and comparative fuel consumption with gasoline:

Fuel	*Consumption Ratio*	
Gasoline	100.0	(Base)
Ethanol	161.4	Volume to equal gasoline
Methanol	221.6	Volume to equal gasoline

If we took these data at face value, methanol would have to cost less than 100/221 times gasoline price to be competitive. For example, if gasoline is at 70 cents per gallon, methanol would have to be below 31.6 cents per gallon.

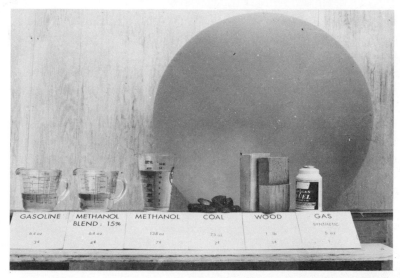

This photograph shows the amounts of several fuels required to propel
for one mile a car that goes 20 miles on a gallon of gasoline. Quantities
were based on calculation or experience, or both. Costs were based on
65 cents per gallon for gasoline and methanol, $50 a ton for coal, $30
for a 90-cubic foot cord of wood. The low cost per mile for the last two
fuels does not consider the high first cost of on-board gasogens for their
use. The gas bottle anticipates a synthetic fuel of high heat content,
such as propane, with 21,670 BTU per pound.

Sphere: 4 ft. 2 in. diameter, holds 40 cu. ft. of "producer gas" (CO
27%, H2 14%). This gas has a low heat content (155 BTU/cu. ft.) and
is shown at atmospheric pressure. An equivalent quantity of this type
of gas flows through a gasogen every mile traveled.

Although this price is considered reasonable by some, even at
present, we cannot expect cars that burn pure methanol for
some time. Hence, an example using blended fuel is more
pertinent.

Assume a 10 percent blend of methanol:

Cost of 10 gallons of gasoline @ $ .60/gallon $6.00
Cost of 9 gallons of gasoline @ $ .60/gallon $5.40
Cost of 1 gallon of Methanol @ $1.00/gallon $1.00
                                    Difference for tankful $0.40

Assume that the car goes 30 mpg on gasoline, and has a 6 percent increase in mileage on the blend, giving 21.2 mpg (a conservative figure). On the blended tankful, the car goes 212 miles, against 200 on gasoline. Twelve miles gives a credit of 36 cents so the cost of the blend is 40 minus 36 = 4 cents per tankful. How does this example seem to cheat the mathematical and thermodynamic analysis in the first table?

The higher latent heat of alcohols reduces intake temperature of the fuel-air mixture, compared to gasoline. Hence there is higher volumetric efficiency for any given horsepower output, for blends up to 20 percent alcohol. After this point, richer alcohol contents increase fuel consumption at rates that are more in proportion to the heat content.

This nonlinear effect of alcohol additives has made possible the surprising mileage improvements reported by Reed and Lerner and has explained reductions in toxic emissions far beyond the 10 to 20 percent alcohol contents involved. The message of the apparent phenomenon is simple and heartening: The greatest value of methanol, as a transitional additive and extender of gasoline, occurs in the range of proportions that requires no change, and little adjustment, to present cars. These facts were established first by E. Hubendick* and reported to the World Power Conference in 1928.

The Principles of Motor Fuel Preparation and Application, by Alfred W. Nash and Donald A. Howes. New York: John Wiley & Sons, Inc., 1935. Vol. I, "Alcohol Fuels," pp. 349-451; vol. II, "Motor Fuel Specifications."

3

Portable
Gas Generators

Soon after the peak of the petroleum shortage of 1973, many vehicles essential for maintenance tasks in municipal services, particularly gas companies' vans, appeared with bumper stickers proclaiming that they were "running on clean fuel." They were equipped with carburetor adapters and accessories enabling them to use a steel bottle of natural gas as their primary fuel source. Except to a perceptive driver, the conversion made no difference in performance of the vehicle and no external, visible change in a car's profile. However, the exhaust is so harmless (water vapor and carbon dioxide) that trucks operating entirely inside warehouses and factories may be so powered.

Facts About
Portable Gas Engines

An infinitesimal fraction of the motoring public is aware of the use of gas as a motor fuel, and of its compatibility with the gasoline engine; accordingly the portable gas-maker has been a total stranger in the United States since 1914. However, it provides a simple, relatively inexpensive means of converting any number of solid fuels to a useful automotive gas, without recourse to the traditional monopolistic systems that produce and dispense petroleum fuels. It makes this fuel while traveling, more or less in proportion to the needs of the engine; it can function on coal, coke, briquets, wood (either branches, chips, or damp waste), or corn cobs, if nothing else is available.

Less Power/More Care

In exchange for this ready versatility the gas generator (gasogen) demands certain sacrifices, the most noticeable of which is diminished power. Even the smaller U. S. compact cars, however, have considerably more power than is considered necessary by unbiased engineers and European economy-car builders.

The second demand, which may only assert itself when it is too late to heed it, is the need for frequent personal attention to renewing filters, to cleaning the purifier system, and to disposing of corrosive and foul-smelling by-products before they have a chance to enter the sensitive confines of the engine's lubrication system.

The loss of power, compared with gasoline or natural gas, occurs because the gasogen uses air to oxidize its carbon fuel, and air is 80 percent nitrogen, which is passed on to the engine as a useless diluter of the real fuel, carbon monoxide. The actual power loss is around 40 percent.

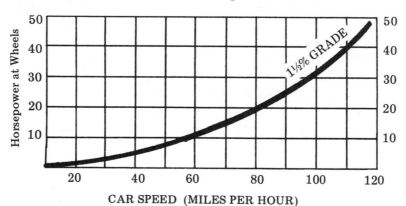

Horsepower required to propel a 1,700-pound, well-streamlined car of a drag coefficient of 0.32 at various speeds. Data computed and experienced. Modern radial tires would improve performance.

There are known techniques for dealing with both of these disadvantages of on-board fuel generators, but it is obvious that the seriousness of world events necessary to drive a significant segment of vehicle fueling to gasogens would also raise the drivers' threshold of tolerance to inconvenience.

The major difficulty, in the event of emergency, would be the immediate acquisition of fuel sources for urban motorists. In the long run, however, the traditional solid fuel sources, coal and wood, provide potential resources for nearly all transportation needs of the foreseeable future, whether produced in transit or converted, with more efficiency, to liquid fuels in large plants.

Heated, Then Cooled

We have mentioned two energy losses in gasogens: nitrogen dilution and the human energy expended in cleaning operations. A third, not as responsive to mechanical remedies, is inherent in the outwardly simple conversion of carbonaceous fuels to combustible gases. This is the loss of heat units occurring when the fuel is burned in the retort, and then cooled before entering an internal combustion engine. The cooling is necessary to clear the gas, by condensation, of acids and tars harmful to the engine, and for volumetric efficiency, as explained later in this chapter.

Although it would be an example of "out of the frying pan, into the fire," this problem could be evaded by changing the engine to an external combustion affair: steam, Stirling cycle, or a gas turbine in which all available heat units are recycled (see chapter 4).

Gasogen Reactions

Reactions taking place within the gasogen are well known and should be understood by everyone contemplating construction and operation of one (see illustration) for reasons of safety, and efficiency in fuel consumption. Although the various zones of chemical activity and temperature were clearly defined for steady state production for city gas supply in manuals of the nineteenth century, there may be wide deviations from the norm in automotive application. For example, the temperature of the combustion zone influences the gas quality drastically, as shown in the accompanying table.

Effect of Combustion Zone Temperature on Gas Quality

Temp. Deg. C.	Temp. Deg. F.	Composition of Gas by Volume			Water Vapor	
		H_2	CO	CO_2	Dissociated	H_2O
674	1247	65.2	4.9	28.9	8.8	91.2
861	1570	59.9	18.1	21.8	48.2	51.8
1125	2057	50.9	48.5	0.6	99.4	0.6

G. *Rouyer, Dunod, Paris, 1938, "Etude des Gazogenes Portatifs." (Quality of fuel gas as controlled by temperature of gasogen combustion zone.)

The second most important determinant of gas quality is the fuel itself, if one is to heed the experiments of one Bell, quoted in the earlier Rouyer reference:

"The following figures show the results of passing a stream of CO_2 over three gasogen fuels, all at identical temperatures. The superiority of charcoal as a producer of combustible gas is evidently due to porosity, and suggests the greater efficiency of pelletized fuel and fluidized bed techniques in the future:"

Gas	Dense Coke	Porous Coke	Charcoal
CO_2	94.56%	69.81%	35.3%
CO	5.44%	30.19%	64.7%

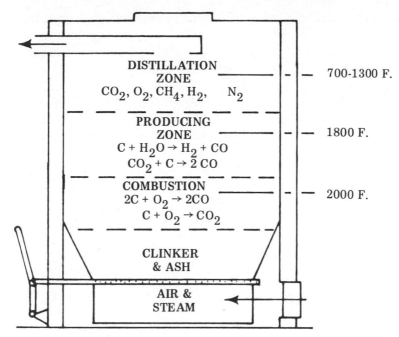

DISTILLATION
ZONE ———————— — 700-1300 F.
$CO_2, O_2, CH_4, H_2,$ N_2

PRODUCING
ZONE ———————— — 1800 F.
$C + H_2O \rightarrow H_2 + CO$
$CO_2 + C \rightarrow 2\,CO$

COMBUSTION ———————— — 2000 F.
$2C + O_2 \rightarrow 2CO$
$C + O_2 \rightarrow CO_2$

CLINKER
& ASH

AIR &
STEAM

Diagram of a nineteenth-century gas producer, showing approximate zones and temperatures. Coal is used as fuel.

Here the superiority of charcoal over coke is evident, with higher carbon monoxide. Bell has neglected to list hydrogen content, which might have made a slight difference. Although the data are sketchy, we have inherited enough from the 1940's to know that the secret of efficient gasogen operation lies in maintaining a constant load on it (speed of engine) or in automatic and highly sensitive control of temperature, which must, in present designs, be linked directly to combustion air input, and, in turn, to intake manifold vacuum and engine revolutions per minute.

Fortunately for several hundreds of thousands of Europeans during the decade of World War II, gasogens on trucks, cars and tractors performed their tasks despite careless operation and maintenance, and despite extremes of temperature and fuel quality. Much of this adaptability and tolerance is due to the simplicity of the basic generator design.

It's Like a Stove

It is no more than a stove, usually turned upside down, and in its elemental form, the gasogen has no moving parts. At worst, it has a moving grate, and an electric or hand-driven blower to get it started. At the top there is a fuel loading door; in about the middle, there is a small door to light the fire; at the bottom another door, for removal of ashes. Thus

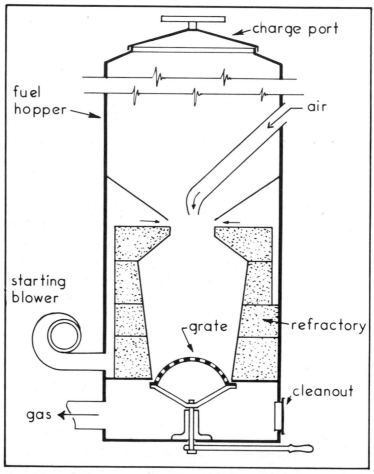

An early and heavy gasogen design.

far the description matches that of an ordinary old kitchen or laundry coal stove.

The flue, or chimney, is what distinguishes the gasogen from a stove, as it is only four inches or less in diameter, as compared to six for a coal stove. Another difference is that air is admitted *over* the fire, which burns downward, side-wise, or upward, depending on the model. Air is drawn in through small tubes to the center of the fire box, instead of through an opening in a door. The tubes are called *tuyeres,* or pipes, as the French claim credit for this piece of hardware, originally for blast furnaces. However, there is no fireside glow about these pipes, as they are shrouded by an air duct, which directs cool air over the inner tips of the pipes to prevent them from melting, and often connects with a blower, used in starting the fire. A pipe leading directly upward, with a shut-off valve, vents the useless smoke of starting the fire to the atmosphere. Illustration shows typical construction for a gasogen.

In the days of coal heating of tenements, the first cold snap of winter brought death to many immigrant families, often from the Mediterranean, huddled in a room with a stove. With windows caulked with newspapers, the air supply to the fuel was inadequate, and a change in wind direction or the addition of coal caused the draft to subside, and carbon dioxide (normal flue gas) passed downward, over the hot coal, losing an oxygen molecule, and escaping into the room as carbon monoxide — bad news in a stove but the life-blood of an internal combustion engine, whose unhappy lot it is to be weaned on a gasogen. This gruesome explanation notwith-standing, the hazards of the gasogen are almost negligible, as it must be airtight to work, and the engine inhales the deadly fumes and emits an extremely clean exhaust of carbon dioxide and water vapor.

Gasogen Design: Trial And Error

The design of gasogens, as may be guessed, is by trial and error, and the major variables are usually controllable from the driver's position. For a certain engine at a given cruising speed, with a fuel of known moisture content, size, heating value and density in the firebox (which is invariably combined with an overhead magazine of fuel), there will be an optimum draft, or negative pressure, in the firebox, adjusted by a valve. As the CO is sucked toward the engine, air for its combustion must be admitted near the intake manifold. This is adjustment number two. The third one is the valve that connects the normal gasoline carburetor to the engine for starting, and then, when the gasogen's fire is going (a matter of skill and minutes) switches off the gasoline and connects the engine with the gasogen.

Think Ahead

It may be seen that there is plenty of scope for skilled manipulations of these controls, as drivers prepare for a long hill ahead, building up a reserve of heated firebed by shifting down, which increases air intake volume. On the other hand, a long down-grade might allow the fire to go out, unless the engine is "gunned" occasionally. One can now appreciate the relative elegance of modern carburetors, that make all these judgments and adjustments automatically.

The largest part of the gasogen's extensive installation is not the retort but the cleaning and cooling equipment necessary to remove tars and acids, (a whole chemistry laboratory-full, if you are indiscreet enough to burn soft coal) and other solids that would contaminate and abrade the engine. French gasogens described in the historical sketch (chapter 5) were designed to run on lignite, imposing maximum cleaning problems. Charcoal and coke, having been cooked until most of their volatile contents have departed, need less attention to cleaning than other fuels. However, their activity and temperature of burning are high, and a large cooling area must be considered.

If the internal combustion engine is fed on air that is hot (i.e., expanded), it cannot gulp as much of it as if the air were cool. The gas from a gasogen is dilute enough — with

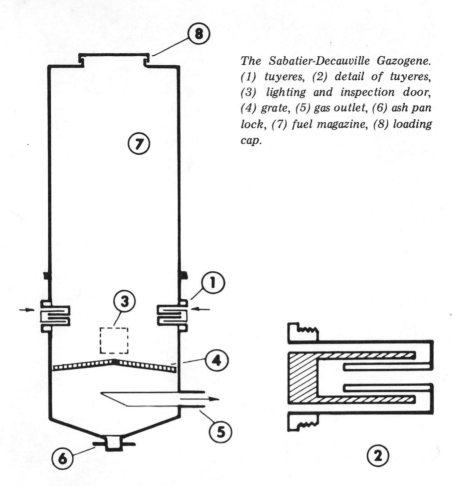

The Sabatier-Decauville Gazogene. (1) tuyeres, (2) detail of tuyeres, (3) lighting and inspection door, (4) grate, (5) gas outlet, (6) ash pan lock, (7) fuel magazine, (8) loading cap.

carbon dioxide and nitrogen — without further loss by heated air intake. This is one reason for coolers of good size (volumetric efficiency). The other reason, equally important, is that the lower the temperature, the less tar gets into the engine, and the more of it sticks in the sumps of filters and collectors.

Many Filter Materials

A list of the materials that have been used for filter beds would fill a page. Some materials, like ceramic rings, stone chips, and metal gauze, are permanent, and must be washed off periodically. Others, like mineral wool, jute or hemp fibers, and wood chips, are disposable (better be cautious). These necessary, frequent service jobs are the real complaint against gasogens; and in a well-designed outfit, quick-release latches and accessible replacement filters will be apparent. The British Ministry of Transport designs for its equipment were excellent in this respect, as they imitated vacuum cleaner fittings that any child can manipulate.

The only part of a gasogen outfit not readily made by the simplest sheet metal and welding techniques is the changeover valve, or carburetor shunt. It is similar to that supplied by dealers in propane conversions, but much larger, as propane gas is *all* fuel and rather concentrated. Another essential, easily purchased, is a gasoline shut-off valve, with electric actuation. This should be switched to the closed position automatically when the gas valve is opened, the carburetor shunted, and the engine begins to run on carbon monoxide gas.

Ferguson tractor with Svedlund wood gas generator.

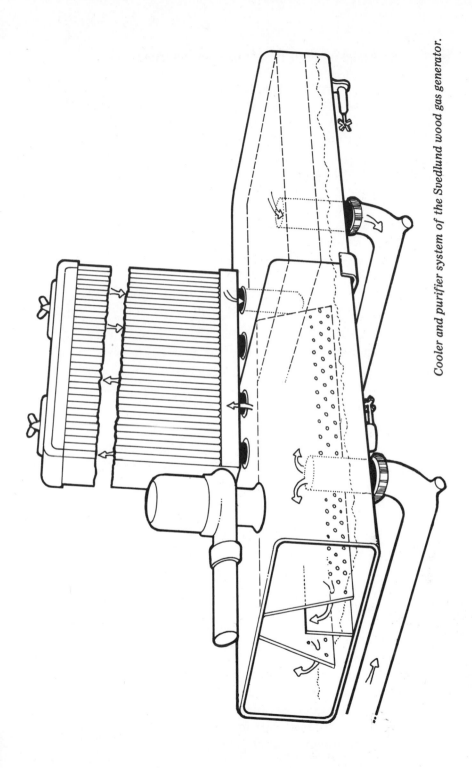

Cooler and purifier system of the Svedlund wood gas generator.

When the supply of gasoline is strictly curtailed, or when it is at such a distance from the user's base that it is inconvenient or expensive to fill one's tank, the annoyances of the gasogen diminish. They are also reduced if fleets of vehicles are equipped with gasogens and maintenance is done routinely by specialists.

Sports cars and racing cars, particularly in European practice, used the positive displacement "blower" or supercharger to increase the amount of fuel-air mixture entering the cylinders. Increases in maximum horsepower and acceleration performance varied, but they might rise as high as 25 percent so long as the increased compression ratio did not cause excessive knock. Modern engines are pushing the limit closely, and supercharging thus seems an unprofitable device for increasing the power of generated gas.

Addition of Oxygen

Oxygen, once an expensive industrial luxury for certain processes where nitrogen is undesirable, is now sufficiently low-priced to be considered for the combustion of subterranean coal and of municipal wastes. A steel bottle of oxygen, piped through pressure regulators to the gasogen and supplied for acceleration and hill-climbing by automatic barometric sensors in the intake manifold, might restore a large part of the power lost through nitrogen (air) dilution, excess carbon dioxide, and low temperature of the reaction zones.

The reactions involved become more favorable to fuel quality as the temperature is elevated; hence, the control of this parameter, as well as internal pressure (normally negative) would surely improve performance. Solid-state electronic controls make this kind of improvement, similar to fuel injection systems, quite inexpensive and reliable.

Performance figures supplied by System Svedlund AB, for equipment redesigned and tested in prototypes in very recent years, are shown in the following table.

Comparison of Fuel Consumption
in Svedlund Gasogen

Fuel	Water Content	Equivalent in Gasoline
2.8 kg. wood (6.1 lb.)	18%	1 liter
23.3 lb. wood	18%	1 U. S. gallon
1.4 kg. charcoal (3.08 lb.)	7%	1 liter
11.65 lb. charcoal	7%	1 U. S. gallon

The Saab, Model 99 L, 1974 (shown on p. 55) is equipped with the smaller Svedlund charcoal system. Its consumption is 20 kg. of charcoal (44 lb.) for 100-110 kilometers (62.14-68.35 miles) at a speed of 90 km/hr (56 mph).

Svedlund made two different systems, for charcoal and wood. Wood gas producers for trucks, buses and tractors cater to engines of 2.5 to 11 liters displacement. Five sizes are listed. For passenger cars from 1.2 to 5 liters, three sizes were provided.

Svedlund, in Orebro, Sweden, still manufactures these systems on order.

Analysis of Gas (Svedlund)

Gas	Percentage, Wood Fuel	Percentage, Charcoal
Carbon monoxide	19	30
Hydrogen	18	7
Methane	1.6	0.5
Oxygen	0.6	0.5
Carbon dioxide	12	1.5
Nitrogen	48.8	60.5
Calorific values	1206 kcal/meter cube.	1228 kcal/cu. meter

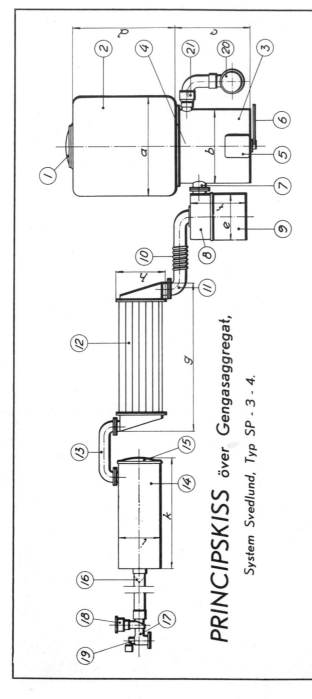

PRINCIPSKISS över Gengasaggregat,

System Svedlund, Typ SP - 3 - 4.

Charcoal gas generator, System Svedlund. (1) fuel door, (2) fuel bin, (3) gas producer, (4) inspection door, (5) ash pit door, (6) shaker lever, (7) gas outlet, (8) cyclone purifier, (9) soot bin, (10) expansion tube, (11) gas riser, (12) cooler, (13) cool gas tube, (14) gas purifier, (15) clean-out door, (16) flame arrester, (17) mixing tube, (18) air cleaner, (19) carburetor, (20) electric blower, (21) check valve.

54

Saab, model 99L, 1974, equipped with the smaller Svedlund charcoal system. These components are made on order by Svedlund, in Orebro, Sweden.

Because the charcoal industry in the United States has deteriorated to a small, almost invisible regional craft, catering to restaurants and gourmet customers, it seems logical to stress wood gas producers, despite the somewhat more tiresome purification problems of this ubiquitous fuel. When and if the superiority of charcoal fuel becomes generally known, charcoal production may be revived on a scale approaching its nineteenth century status.

Constant-Speed Operation

The gasogen, with any fuel, is not ideally suited to the stop-go driving characteristic of family suburban driving. Its best economy is achieved in constant-speed operation, such as cross-country trucking, stationary farm generators, pumps, mills, and so on. Marine engines also are ideally suited for gasogens, as demonstrated by past history (chapter 5). However, drivers who log over 10,000 miles a year may well consider the economies available using wood gas.

Costs

Rough predictions of the amortization of equipment can be made with the chart shown here, constructed from the following data: Gasoline consumption of the car is assumed to be 20 miles per gallon. At 10,000 miles, it will have consumed 500 gallons. At 60 cents per gallon, this comes to $300. Another line shows the increased slope due to a price of 80 cents a gallon.

A wood gas generator is assumed to cost $500, on a mass-production basis. Common North American woods, their weights, and relative heat values are given below, all at 12 percent moisture content. A legal cord contains a net wood volume of 90 cubic feet.

Consumption of wood in a gasogen is variable between much wider limits than gasoline consumption. Hence the last column is based on an average of one mile per pound of wood. On charcoal, this figure may go up to 1.4 miles per pound, according to data supplied by Svedlund for a model 99L Saab, at 56 miles per hour (90km/hr).

Heat Values of North American Woods

Wood	Weight, Cord	Heat Value, Millions BTU/Cord	10,000 Mile Estimated Consumption, Cords
Birch Maple	3,960 lbs.	30	2.50
Oak	4,230	32	2.37
Pine	2,250	18	4.54

It Pays For Itself

It is apparent from the graph that, with current (mid-1975) gasoline prices and high-priced dried cordwood, the gasogen may pay for itself at about 34,000 miles, but with hard wood at $30 per cord and a gasoline price of 80 cents per gallon, the break-even point becomes 15,000 miles.

Cost of gasogens vs. *operating costs.*

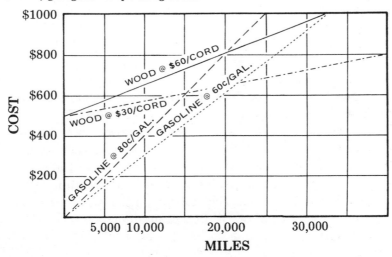

4

Alternative Engines and Fuels

While the storms over energy supplies have centered on petroleum and specifically gasoline and its champion consumer, the ordinary internal combustion-engined car, a slight squall arose around the external combustion engine in the mid 1960s. The steam engine is "externally" fired; the fuel either gasoline or fuel oil, is burned at atmospheric pressure outside of a boiler, instead of in the compressed atmosphere of the cylinder of a gas engine. The result of the former is that the exhaust lacks most of the pollutants that the Environmental Protection Agency is struggling to eliminate from our environment. The hot-air, or Stirling engine, is in the same category.

The Steam Automobile

Although the steam automobile existed in great numbers and performed creditably in the decade 1900-1910, it used almost the same fuel as its noisy successor, the internal combustion car, i.e., gasoline and kerosene. The "revivals" that have occurred since that era have been statistically tiny and depressing, from an engineering viewpoint, although the popular scientific press has capitalized on the legend of steam cars and the eccentricities of their promoters.

The last revival, actually the final death-twitching of an exhausted subject, was extended by the emissions control movement. To revive a romantic vehicle of the past to meet the clean-air requirements of another era became an exciting crusade for a hundred inventors and thousands of newspaper scribblers. In the momentum of actual prototype construction of cars and buses, it was barely noticed that these steam vehicles were gulping more fuel than their internal combustion competitors — as much as double, in fact. If the Arab oil embargo had not occurred, the twitching might have extended longer.

Despite the flimsy appearance of such cars as this 1902 model, ancient steamers had good performance. This 500-pound model climbed Mt. Washington and Pike's Peak. Models built after 1910 became too complex, too heavy, and too expensive to remain in competition.

Steamers Are Easier to Operate

The legend of the steamer's superiority started in the first decade of the century, with the French Serpollet steamer, the Locomobile, the White, the Stanley, and other lesser brands made in England and the United States. At this stage of development, the internal combustion- or "explosion"-engined-cars suffered several disadvantages. They needed to be cranked. They required priming in cold weather, and more cranking. Gearshifting was a losing gamble. They stalled easily, and their safe management was an art that men would not often entrust to women. The steamer, by contrast, could be fired by lighting a wick. It started, sometime later, by pressure of a small, white-gloved finger on the hand throttle, and it could not be stalled. There were no gears to be shifted, no clutch, and no noise. They were dependable enough to be preferred by many medical doctors.

This Stanley Steamer roadster was modernized with Space Age materials.

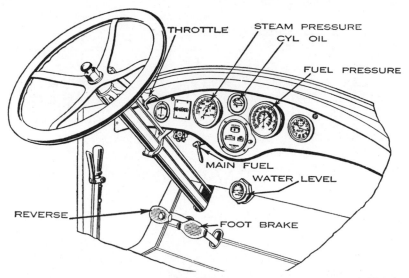

THROTTLE   STEAM PRESSURE
CYL OIL
FUEL PRESSURE
MAIN FUEL
WATER LEVEL
REVERSE
FOOT BRAKE

The controls of the Stanley Steamer were simple and functional. This is a diagram of the interior of the Model 740 Stanley. The car weighed 3,825 pounds and was priced at $2,750 for the seven-passenger open car, the five-passenger phaeton and the two-passenger roadster. The five-passenger sedan cost $3,585 and the seven-passenger sedan cost $3,985.

In the next decade, the self-starter was perfected for gasoline cars, gears became more docile, and spark ignition became more reliable. The steamer lost ground by comparison, and only the Stanley survived the World War I years, limping along on 600 cars a year and the patronage of a diminishing band of supporters, into the middle of the postwar decade, with the Doble flashing briefly across the stage. The once-superior performance of the steamer, its smooth acceleration and indomitable hill-climbing ability, was finally eclipsed by the gasoline engine and automatic transmission.

Steamers Costly

Good steamers were never inexpensive. The large White steamer sold for about $5,000 at a time when the equivalent Rolls-Royce sold for the same amount. In the 1920s, the

THE FASTEST CAR IN THE WORLD
(Rate of 127.66 Miles an Hour)

In 1906 this Stanley Steam Car was called "the fastest car in the world, after achieving a speed of 127.66 miles an hour. (The Bettmann Archive)

A charming holdover from the horse carriage days was the whip socket (shown near driver's left hand) on this early Locomobile steam car. In this 1905 photograph, driver is getting water from water spout. A few cups would not take this steam car far, since it had no condenser to circulate and reuse water. (The Bettmann Archive)

Doble was priced from $20,000 upwards, depending on the owner's tastes in coachwork. The Doble was a huge car, with one model weighing 6,000 pounds and having a 151 in. wheelbase (see illustration).

To be reasonably efficient, the steam engine must operate at high temperature, and that calls for expensive nickel steels and the difficult machining and assembly of alloys. Only one part of the internal combustion engine — the exhaust pipe — works at a temperature endured by almost the entire boiler and engine of steam plants. Some of the practical problems, such as temperature, lubrication, and routine maintenance, might be solved in production by new materials and techniques, but the basic thermodynamic handicap of steamers cannot be overcome by any breakthrough. (See p. 67)

Lear's Experiments

High hopes and fat wallets have set aside the thermodynamic facts, if not overcome them. William Lear, who was happily and profitably building jet airplanes in 1960, decided to try his hand at steam car design. With encouragement from

This 1925 Doble Steam Car had a 151 inch wheelbase and weighed 6,000 pounds.

ecologically-motivated fans, from Senator Muskie to the kookiest of inventors, he set up shop in April, 1968, to build a prototype steam car and a race car for the 1969 Indianapolis 500-mile race. There was talk of testing a car for the Highway Patrol (State Police) in California, and of a fluid to replace water as the power medium, officially called *Learium,* but dubbed *D-Learium* by steam engineers. The man who pioneered the car radio and developed a successful small jet and an automatic pilot was going to extend his magic touch to the legendary steam car.

Several miscalculations and two years later, not much automobile appeared for the $10 million Lear says he spent on development. A "vapordyne" car, numbered 23, with a complicated engine and an invisible boiler, toured the auto shows of 1969, behind ropes. Besides a Neanderthal disregard for the laws of heat transfer, the Lear venture foundered on poor judgment — a decision to build a 300 horsepower engine — and on misplaced faith in the ability of employes to design in fields in which they were untrained.

Left: *Steam power isn't confined to automobiles. Wilbur Chapman is shown here with a 300-pound steam outboard engine he designed in 1947. Boiler added another 200 pounds of weight to this power plant.*

Right: *The desire for a "modern" steam car drove many mechanics to cannibalize old Stanleys for their engines and boilers. Here is a 1945 special that was assembled by Harry W. McGee (in foreground), using a 1925 Stanley engine.*

Buses in California

More significant projects were afoot in 1969. The State of
California, where Los Angeles County is the most densely
automobile-laden spot on earth (over 3½ million vehicles at
that time), requested proposals for supplying four steam buses
for testing. While one-half million dollars was allocated,
technologists called for $60 million to develop clean power

*This early American automobile was capable of eight miles an hour. It
was first shown at the 1892 World's Fair. In this photo, taken in Los
Angeles, the racing driver Ralph de Palma is signalling, while engineer
Joseph Wright handles the controls. (The Bettmann Archive)*

plants. To make a long story short (every steam story is long), a General Motors bus was running in mid-1972, powered by a Lear steam turbine, with tap-water as the fluid. Although the level of emissions was clearly below the limits of the 1975 standard, fuel consumption was nearly a pound per horsepower per hour — over double that of the diesel engine replaced.

When they responded to the pressures leading to investigation of alternative power plants, established automobile manufacturers in the United States made similar mistakes in transferring gasoline engine horsepower traditions into steam engine requirements. While it is possible for a limited number of affluent customers to support the fuel bills of a 300-horsepower, three-ton gasoline car, a steam plant of this power pushes the physical limits of space available and exceeds, by about 50 percent, the monstrous fuel consumption of the "prestige" and sports cars.

This steam powered car was designed and built by Skip Hedrich in 1969. (World Wide Photos)

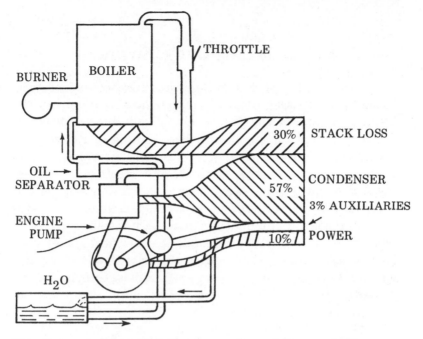

This diagram indicates that the steam car's reputation was built on performance, not efficiency. It was the least efficient of all motor cars. Available power was 10 percent of the heat content of the fuel used.

Saab Project

A refreshingly different approach is being taken by Saab, the Swedish aircraft and car maker. Saab's experimental car, seen by the press early in 1975, has a small engine of fully balanced design and a boiler and control system using latest technology. One of Saab's goals, a light-weight boiler (28 lbs.), is achieved by the use of tubing the size of soda straws.

Although there will be no "breakthroughs," the experience Saab gains may give that firm a head start in the event of rapid shifts in fuel economics and synthesis. The steam engine, in addition to its capability of handling great overloads, can use any gaseous or liquid fuel, without regard for octane numbers, emissions control accessories or noise problems.

The Stirling-Cycle Engine

The revival of the Stirling-cycle engine has been unspectacular and uninteresting to all but an engineering minority. The idea of an engine using air instead of steam to drive its piston gained a patent for the Rev. Robert Stirling of Kilmarnock, Scotland, on Nov. 6, 1816. After a modest success as a power plant for butter churns, grinders, and small manufactures until the gas engine came into use, the hot-air engine descended to the status of a toy. It reappeared about 1938 as a silent generator component for military installations. N. V. Philips Co., a Netherlands research and electrical manufacturing concern, improved the efficiency (which was very poor) and reduced the size of the ancient models. General Motors looked into the system for space ships and cars but abandoned it about 1970.

The Ford Motor Co. is now working under a Philips license on an unusual car engine in which air is replaced by

Author John W. Lincoln with a working model of the Stirling engine.

hydrogen as the working medium. Long ago it was recognized that air had a poor capacity to transfer heat, compared with steam and other gases, particularly those called "monatomic." But air is cheap and harmless, and the original Stirling plants were safe, if not efficient. Modern engine designers have at their disposal certain sealing techniques that permit the retention of an initial charge of an elusive gas like hydrogen or helium for the life of the engine. The Philips-Ford engine provides speed changes by quick changes in the pressure maintained on a reservoir of the gas. Without this device, heat generated by the fuel burner would be transferred much too slowly through the cylinder walls to accelerate a vehicle in traffic. Conversely, a drop in reservoir pressure makes the gas far less conductive of heat, and causes deceleration.

More efficient than any other external combustion engine, including gas turbines, the Ford-Stirling engine is estimated to cost over $3,500. Starting time is long, like the steam system, but there is no problem of freezing and no water and oil separation problems. As in a steam system, the external combustion of fuel leads to very low emissions. It is the only type of engine that is both nonpolluting and capable of considerably higher efficiencies than internal combustion engines now in use. This potential derives from the Stirling's facility for regeneration or recycling of heat from one cylinder to power another, a basic thermodynamic advantage that has been recognized by the Jet Propulsion Laboratory's request for grant funds to develop it, along with turbines, in a 10-year, billion-dollar program.

The basic Stirling engine is remarkably silent. Even with such a high price on this commodity, it is difficult to foresee a commercial automotive use for the system, except for a deluxe limousine. Since the usual fuel for both steam and Stirling engines is petroleum, either kerosene or jet fuel, there is little point in looking to them as oil-savers. However, with a cheap source of non-oil fuel, such as wood gas, methanol, or ethanol, these external combustion systems might become economical, especially where the acoustic environment of a hospital or resort area demanded near-silent power supplies.

Ryder-Ericson hot air engine, built about 1840-50, in the New England Steam Museum, East Greenwich, R. I.

Gas Turbine Engine

The automotive gas turbine, pioneered by Rover in England, and by Chrysler in the United States (1960's) has been plagued by a high count of nitrogen oxides in its exhaust, an inherent feature of combustion under compression. The Environmental Protection Agency has awarded Chrysler a contract to seek improvements in this emission problem. Turbines required to propel vehicles at widely varying speeds must necessarily link to driving wheels through high-ratio gearing or electric drives. Hence, the turbine, because of noise, cost of gears, and cost of high-temperature alloys in the engine, is not a likely candidate for replacing the internal combustion engine.

Another impediment to the turbine's use in cars, at present, is its need for petroleum distillate fuel in somewhat larger amounts than luxury vehicles can tolerate. Large marine and stationary turbines may be very efficient, and some of the latter have run on powdered coal. Probably large gas turbines will be coupled with coal or rubbish-fired gasogens in the future, but the small automotive plant seems remote.

Electric Cars

Compared with the extensive connoisseurship of antique steam cars, there is very little popular enthusiasm for the electric, a vehicle associated with little old ladies, shaded urban streets in Boston, Brookline and New York, and, for very good reasons, the level landscapes of many Midwestern cities. Its advantages in the first quarter of this century remain much the same today: cleanliness. No exhaust, no dripping fuel or lubricant. No noise. Instant starting, with finger-tip control, and gearless speed changing. Electric cars were, and still are, ridiculously cheap to operate; the nocturnal charge costs a few dimes. They need no tuning, no oil changes, no cooling fluids.

There are several hundred thousands of electric vehicles in operation now, all over the world, but most are commercial trucks, vans, golf carts, etc. England is estimated to have 50,000 in service. Many of those in the United States are running inside shops, warehouses and industrial plants, and on private property and therefore are not registered or counted in usual statistical summaries. While commercial users may choose electrics for their specific advantages, the motorist commuter or salesman is barred from the choice of electrics by their obvious disadvantages on the public highway. These are shown in the following list.

1. *Short range, about 20 miles.* This means 20 miles out and return, a more realistic definition than saying 40 miles total.

2. *Poor performance.* Although no electric owner expects jackrabbit starts, he may be embarrassed by toots at the traffic lights and the need to keep near the curb in heavy, faster traffic.

3. *Slow speed.* Although not a necessary limitation, it is a trade-off for range. The electric that can maintain high speed and long range does not exist.

4. *Initial cost.* Electric motors, switch-gear, and especially batteries are all made of fairly expensive metals.

The author's 1914 Rauch and Lang Electric Brougham. Modern electrics have better brakes, but far less comfort. The power plant has not changed.

Refueling the Citicar *is a matter of plugging in the cord. This electric car is manufactured by Sebring-Vanguard Inc., Sebring, Fla. It has a 48-volt system with a top speed of 38 mph, and the manufacturer lists a range of "up to 50 miles."*

Fossil Fuels Needed

The electric, like the steam car, if it displaced the traditional gasoline car, would solve the air pollution problem pertaining to vehicles, but neither can do anything for the energy problem. Fossil fuels are needed to keep generating plants going, and a significant number of electric vehicles on charge at night, even during "off-peak-load" hours, would erase these slack periods and demand extended plant capacities — exactly what the utilities cannot stand, now or in the near future.

However, when and if these utilities return to the burning of coal, under the refined circumstances of scrubbed and purified exhaust gases, that plentiful domestic fuel can be burned more reasonably and cleanly in huge boilers than can oil in individual vehicles. The electric, although it is unlikely to be as efficient as a directly-fueled heat engine, may thus justify its existence on the basis of clean air and silence, in certain environments.

Cars Being Built

There are several electric passenger car projects at the time of writing, and a growing production line count indicates that several dozen customers each week are willing to spend about $3,000 on a specialized second car having severe limitations and gratifying rewards. In Florida, Sebring-Vanguard counts its *Citicars* in the 500's, and manufactures about six per day. It is a two-passenger, squat affair using eight conventional lead-acid batteries. Its chassis is basically that of a golf cart, with a simple plastic sheet body on a tubular frame. Curtains help keep out the rain.

The second production electric, the *Elcar,* comes to the United States from the Italian shops of Zagato, a well-known coach-builder of sports cars, via Elkhart, Ind. Although the *Elcar* is of better appearance than the *Citicar,* the two are similar in power: 3 to 3½ h.p.; batteries, 48 volts, lead-acid; maximum speed, 25 mph.

Also in Italy, Fiat is developing a small, ambitious electric with a 50 mph. speed and a range of 65 miles. In Japan, where incredibly severe urban pollution encourages government support for electric car research, there are rumors of production models in the near future.

Toyota is already testing an experimental electric with a range of 60 miles, a top speed of 56 mph, and a braking system that is entirely electric, feeding energy back into the battery. Volkswagen and Daimler-Benz also are jointly involved in fundamental research and plans for electric vans.

The Big Three in the United States have all paid lip service to the concept of the electric second car, but none of them went further than the prototype stages. General Motors knocked itself out of any practical competition by choosing a silver-zinc battery with an astronomical price tag. Ford pinned its faith on a battery using sodium in the molten state — one of the most unstable, explosive metals in the

The Yardney Silvercel, an experimental electric automobile. (World Wide Photos 1967)

A closeup of the trunk of a Renault converted by Yardney Electric Co. of New York City to electric power. Batteries replace the engine and fill the luggage compartments.

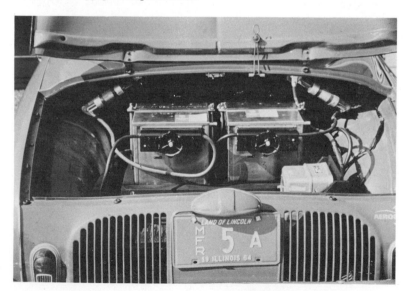

chemistry books. American Motors likewise cast caution to the winds and built a lithium-nickel battery car. Chrysler tinkered with a Simca with twin motors, but a 16-mile range.

Making a Better Battery

Obviously the automobile manufacturers the world over are not competent or willing to push the development of a battery that will give both extended range and lighter weight. Battery manufacturers, however, show more interest and ability. The Yardney Electric Corp., already experienced in developing silver batteries for missiles and space craft, is developing a nickel-zinc battery. Gould, Inc., a large manufacturer of conventional batteries, is working on nickel combinations, as they are not too different from lead-acid in construction and service problems.

    The truly "lightweight" batteries are plagued by cor-

The Islander, manufactured by Electromotion, Bedford, Mass., has a top speed of 28 miles an hour and can carry four persons. Its manufacturer claims a driving range of 40-50 miles. It carries four persons. Its batteries have a life of about 15,000 miles.

rosion, the disadvantages of highly-heated electrolytes, toxicity and high cost. The developers all admit that they are not for this decade. The only possible exception is the zinc-chlorine battery, operating at normal temperatures, with materials that are cheap enough to make it competitive with the standard lead-acid battery. The problems yet to be solved are the containment of lethal chlorine gas in case of an accident, and the relatively short life of about 500 recharging cycles.

Despite the re-emergence of a market for passenger electrics, and enlarged research budgets for new battery development, the old lead-acid style will be with us for years to come. It is not too bad to keep the British, who have always managed to make prudent long-term investments in their own high-quality machinery, from using fleets of electric trucks in all cities. The dairy industry alone uses 11,000 in London, with one chain owning 4,500 vehicles. By comparison, the largest fleet now being assembled in the United States, 350 electric jeeps for the Postal Service, seems a chicken-hearted venture.

Fuel Cells

Battery technology, although it has a history of more than a century, seems to have come to an invisible barrier, forcing scientists to examine the possibilities of converting fuel directly into electrical energy without the need for mechanical and magnetic devices such as boilers, engines and rotating generators. A fuel cell with no moving parts may use a gaseous fuel, hydrogen, hydrazine, or liquid methanol, with oxygen or air as the oxidant. A rather large number of cells is required to boost voltage to useful levels for an automotive motor drive. Although fuel cells are in operation in specialized applications, usually stationary sources of small amounts of electrical current, they are hampered by maintenance problems, and the fact that platinum seems to be the only practical material for electrodes. Hence, the "engine" of a fuel-cell car would cost about $10,000.

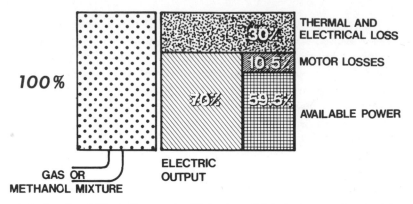

THERMAL AND
ELECTRICAL LOSS

MOTOR LOSSES

AVAILABLE POWER

100%

30%

10.5%

70%

59.5%

GAS OR
METHANOL MIXTURE

ELECTRIC
OUTPUT

A fuel cell vehicle is the most efficient available in terms of output to input. However, astronomical first costs make it a poor practical substitute for the standard product.

Electricity powers this 12-passenger suburban bus, a product of the Battronic Truck Corp., Boyertown, Pa.

There is no noxious exhaust from a fuel cell, the products of the oxidation being carbon dioxide and water vapor. Platinum is being consumed at increasing rates in catalytic converters, and it is unlikely that existing mines could satisfy another demand of any size.

Looking Ahead

The normal internal combustion engine, gasoline and diesel, with devices to control emissions, with basic design improvements such as stratified charge, and with fuel additives such as methanol, is the one sound bet for the general purpose car through the year 2000.

The electric vehicle will grow in numbers, particularly in the form of trucks, as specialized vehicles, and as second cars for local personal transport. Widespread use of methanol blends in the first category would ease present shortages, remove gasoline and oil import problems, and prepare the way for later total replacement of gasoline as a motor fuel.

Steam cars, and those propelled by Stirling engines, fuel cells, and hydrogen will have to wait a long time, either for an improbable technical break-through or an equally improbable change in the economic system, wherein waste becomes a virtue.

Next: The Age of Hydrogen

The fuel of the early internal combustion engine was city gas. The automobile became possible when inventors found a way to vaporize, atomize and burn a liquid fuel: gasoline. A gaseous fuel is still very impractical for vehicles. Nevertheless, many scientists and amateurs have become "hydrogen-ophiles," hoping for the creation of a secondary energy system based on this fuel. The present internal combustion engine will not change over to a hydrogen diet unless a very serious indigestion problem is solved.

Hydrogen is plentiful, it burns cleanly (water vapor is the only exhaust), it is easy to transport, and it is nontoxic. It is being used with high safety scores by NASA and by the Atomic Energy Commission. Over the past 100 years it was used, in a 50 percent dilution, as "coal gas" or city gas. The hazards of this fuel are mainly in the toxicity of the odorless carbon monoxide that makes up a large part of the product. Its disadvantages are in its low heat content (only one-third that of natural gas) and its present cost.

Hydrogen Production

There are at least three ways in which hydrogen can be separated from water, and several ways, in addition to the present internal combustion engine after much modification, that it may be utilized in vehicles. The oldest (1800) and simplest extraction process is *electrolysis,* or the dissociation of water by direct-current electricity. Hydrogen is liberated over the negative pole (cathode), and oxygen over the positive (anode). Although most modern generation of electricity is in the form of alternating current, there are few problems in converting from one to the other. With this system, it is obvious that the cost of hydrogen must always exceed the cost of electricity. However, hydrogen may be stored far more efficiently than electricity and may be transported by many inexpensive means. The theoretical and practical efficiency of the process is high, and the capital costs are within reason — about 45 cents per million BTU of capacity. To be commercially effective, a larger market for oxygen, produced simultaneously and in half the quantity as hydrogen, must be found.

Sites for Power Plants

It may soon be necessary to consider the Arctic and Antarctic zones as the only suitable sites for fossil-fueled and nuclear primary power plants because of present problems of thermal marine pollution as well as particulate atmospheric releases. Also, maximum thermal efficiency is obtained with the lowest possible "sink" temperatures available in unlimited icy oceans. Since these plants would not need to be operated at maximum outputs the year-round (a normal state of affairs), their excess, or peak-load capacity, could be absorbed by hydrogen production with tankers returning the products to populated zones in pressurized containers.

Another method, nonelectrical, for the production of hydrogen is being investigated by a European research team. This is a closed, four-stage *thermo-chemical process* in which heat and water are the only inputs. By eliminating the necessity for electrical power, great savings would be possible. The potential is exciting, as is the prospect for the fuel cell's conversion of gas energy to electrical power without rotating machinery. Both are elegant, philosophically, and elusive, practically.

The third method of hydrogen separation is probably most attractive to solar heat enthusiasts and biologically oriented thinkers. Blue-green algae and other plants are dispersed in cells receiving solar radiation. In the *photosynthetic process,* oxygen is removed from the water, and hydrogen released. Under laboratory conditions, a reasonable efficiency is achieved. The earth receives 0.8 calorie per square centimeter per minute. The photosynthetic cell yields 500 kilogram calories per square meter per day. Based on this, a commercial hydrogen plant (500 tons/day capacity) would need an area of 14,000 acres, or 22 square miles for solar collectors.

Effort is Slow

With that forward look on record, it is sobering to observe the excruciatingly slow pace of past efforts to capture more energy with less effort. In recent decades, windmills, tidal power plants and thermal seawater engines, using the difference in temperature between tropical surface water and adjacent deeps, have been investigated and tested. Systems depending on weather, wind, and sun must be backed up by conventional emergency systems, and intermittently active plants, (tidal and solar) must have storage facilities. Perhaps the advice of a Committee of Common Market countries should be kept before us in future deliberations on energy sources: Watch the growing role for coal about the year 2000. An estimated 87 percent of the world's petroleum reserves, and 73 percent of natural gas will be depleted by that time, but only 2 percent of coal reserves. Coal can be converted to gas and to liquid fuels, but the costs of transportation were such as to delay commercial uses until the 1973 crisis raised the price of oil.

A very quick appraisal of the huge spaces required and the depreciation on enormous capital outlays expected for these solar and natural systems leads, inescapably, to the conclusions reached by studious foresters years ago: Trees are extremely efficient converters of solar energy to carbon fuels. Recently, it appears that they can be further converted to liquid methanol in portable plants.

A Poor Fuel

Hydrogenophiles generally ignore the fact that hydrogen is an abominable fuel for the present high-compression automobile engine. It causes knocking, poor power output, and has a flammability range in air that makes a small leak an enormous hazard.

	Hydrogen	*Butane*
Limits of flammability in air, by volume	4%-74%	2%-8.6%

But when safe and cheap nuclear power becomes a reality, hydrogen may have a chance as a motor fuel, either in a fuel-cell car or in some new form of combustion engine that may be fed hydrogen from storage in the form of liquid or solid metal hydrides, for example.

Some progress in the latter category may come out of research in metal-air batteries, which are regenerated by hydrogen. Cadmium plates and a platinum catalyst raise the initial cost and skeptical eyebrows, but operating costs might be as little as a half a cent a mile. This compares with fleet operation costs for gasoline vehicles (with gasoline at 55 cents per gallon) of about 6 cents per mile, and electric battery truck costs of 6 cents per mile, including battery amortization. If maintenance costs are included in the mileage costs, the gasoline vehicles trail the electric battery cars significantly, and the hydrogen plant is a totally unknown quantity.

Hydrogen Engines

While hydrogen fuel poses difficult problems for conversion of present gasoline engine designs, the trend to higher petroleum prices, along with mounting costs to meet tight emissions standards, has launched research on four hydrogen engines at Oklahoma State University. Forced injection of hydrogen and delayed ignition are claimed to avoid the usual pre-ignition knocks. To provide hydrogen storage in an automobile equal to a 20-gallon gasoline tank, with presently available techniques, would require a 500-lb tank for magnesium hydride. This compound must be heated to produce gaseous hydrogen.

The weight and complexity of the above system, far exceeding that of gasogens of World War II vintage, suggest that the portable generator of "producer gas," a mixture of hydrogen and carbon monoxide, is a more realistic field for progress than hydrogen.

All of the alternatives to the mass-produced internal combustion engine and to its presently accepted fuel, gasoline, described above, have at least one thing in common: They have a popular following, in addition to the respectable scientific and technological communities that have special interests in promoting their more widespread use.

There are, in this popular front, a certain number of crackpots, the "lunatic fringe" of the autophiles, dominated by the buffs of the steam car. At antique car rallies, I have been greeted by several of these as follows: "I am Tom Jones, another steam nut . . ." After recovering my composure, I have listened to their praises of the steam car and their inventions that they feel sure will expedite the practical readoption of the steamer in lieu of the explosion engine.

What About Water?

With varying degrees of forebearance and sympathy, these people can be told a few facts of thermodynamics that will either alienate or educate them. But what does one say to the proposal that one's car will run on water? Here is a crackpot proposal that needs more serious attention than does the revival of steamers.

Although not a fuel, water has returned periodically as an additive to the diet of the gasoline engine. The author's family sedan of the 1920's was equipped with an accessory water jet on the intake manifold, a piece of hose, and a gallon tin can. The gentle spray of vapor — the can was refilled every 20 to 30 miles — quieted the knock of the high compression engine, and seemed to improve performance and economy. Improvements in this old art were reported in the 1974 congressional report. Rather than using the spray intake, the water (18 percent) was mixed with gasoline (82 percent) and emulsified.

Test results showed lowered combustion temperatures and resulting low NO_x emission. The University of Oklahoma has joined in a program of testing this idea with the U. S. Postal Service in Norman, Okla. Part of the fleet there was adjusted for the experiment, supposedly on a continuing project. No cold-weather experience has been reported, and no reduction in economy was observed.

It is annoying for large consumers of gasoline to calculate how much they might have saved over the years had they replaced 18 percent of their fuel with tap water. My family's estimated waste for a 50-year stint of driving without water comes to about $3,000. Such are the gaps between technology and education.

5

Past Uses of
Synthetic Fuels

"Why didn't someone warn us?"

Since the energy crisis and gasoline shortages of 1973, we have heard this many times. The truth is, of course, that we *were* warned — many times. Unfortunately, the history of forecasting the depletion of natural resources has attracted few scholars with loud voices.

In the past similar warnings have been issued — and acted on — in other nations. A look into their histories shows that while these nations may have reacted to the warnings in different ways, nearly all made some positive move to assure themselves of a continuing supply of fuel for transportation.

In the period between world wars, realistic European nations knew that World War I hadn't been the "war to end all wars," and they prepared for yet another. Some nations concluded that a major war would cut them off from supplies of oil. All knew that the successful prosecution of a major war would require an ample alcohol supply both for fuel and for the manufacture of high explosives.

Fuel Blending

The pattern in most of Europe was the same — more and more measures to force the motorist to use a substitute fuel blended with gasoline. By 1937 this program, which sub-

sidized the alcohol industry, cost European governments
$234 million. That year 18 percent of European motor fuel
was produced from wood, coal and agricultural products.
Imported gasolines were selling for 9 cents per gallon before
taxes, but alcohol in 11 countries averaged 44 cents per
gallon.

Alcohol was not the only substitute used. Synthetic
liquid fuels from coal, and gas, generated on board, were
others. Thus, in 1937, European consumption of substitute
fuels was:

Fuel	*Metric Tons, Thousands*
Alcohols	510
Synthetic gasoline, from coal (By Fischer-Tropsch process)	929
Benzene	824
Shale oil	38
Compressed gas & wood gas	234

While the driving public was given little choice in the
matter, being forced to purchase the more expensive blends if
they desired to continue to drive, it should not be inferred
that they were purchasing an inferior fuel, even in the critical
50/50 blend. The blended fuel could provide more power,
reduce or eliminate knocking, and produce a cooler and less
noxious exhaust. It could eliminate the addition of tetraethyl
lead, an advantage not recognized until recently. Its only
disadvantage, hardly noticed by those who could afford to
own automobiles, was its increased consumption compared
with pure gasoline.

The German Experience

In view of the dominant role of Germany at the outbreak of
World War II and her lack of a major oil supply of her own, it

is not surprising that she made the most significant quantitative steps toward an independent fuel supply.

The processes for development of synthetic fuel from coal that were developed in those pre-war years in Germany are the ones that hold the most promise for those in the United States urging use of coal and municipal wastes for this purpose. They are the Lurgi process, now being revived in Frankfort; and the Pott-Broche, the Bergius-Pier, and Fischer-Tropsch processes, all developed prior to the Nazi era. When that shadow fell, a dozen plants were created, and by 1944, they were producing over 4 million tons (32 million barrels) of synthetic oil and gas annually from coal.

With so great a need for fueling a war machine, a military dictatorship did not reckon costs only in marks; thus it would tolerate the process inefficiencies that resulted in expensive fuel in terms of coal consumed. Today, with oil prices rising, Germany is reconsidering a return to her deep coal mines for another go at synthetic fuel production.

While Germany's pre-war production of motor fuels was not impressive when compared with wartime needs, the figures do illustrate the percentage of home-produced fuels, figures that were to increase geometrically through the war years.

German Domestic Fuel Production, 1932

Benzol (from coal)	200,000 tons	44.4%
Alcohol	80,000 tons	17.7%
Gasoline	170,000 tons	37.7%
	450,000 tons	100.0%

This total represented only one-third of Germany's vehicle and aircraft fuel consumption for the year, with the remainder furnished by such friendly nations as Roumania.

The first German government decree on fuel called for the addition of 2.5 percent alcohol to all imported and domestic gasoline. This was later elevated to 10 percent. It was enforced with difficulty, due to inadequate supplies of good-quality methanol. Two common fuel blends, Monopolin and Bevalin, called for 25 percent alcohol and 75 percent gasoline; a third, Aral, was composed of 20 percent each of alcohol and benzol, with the remaining 60 percent gasoline. In the days just prior to the outbreak of war, the nation specified that all motor fuels would be blended from 70 percent gasoline, and 10 percent each of ethyl alcohol, methanol and benzol.

But Germany's enemies-to-be were taking action, too.

France

Soon after the cessation of World War I, France moved to a major domestic grain alcohol fuel industry by establishing a 50/50 mix by volume of ethyl alcohol and gasoline, calling it "carburant national," and decreeing that it be used by government departments and Paris buses. The government required the oil companies to buy the alcohol, to blend it, and to suffer any resulting losses.

In 1931 the law was broadened to require all commercial vehicles to use a new blend, called "heavy carburant national," that used 25-35 percent alcohol. Shortly after passage of that law, another was passed that reached into the tanks of private autos. This law called for the addition of 10 percent alcohol to *all* imported gasoline. Although annual national production of ethyl alcohol was less than 200,000 barrels — France had not yet taken to private wheels — the legislation seemed to have achieved its purposes: to reduce losses in foreign exchange to oil-producing countries and to bolster the domestic ethyl alcohol industries for military exigencies. Complaints were frequent about the legislation, and its result, desultory performance with "carburant national." The half-and-half mix made it quite difficult to start one's vehicle on a cold morning. About 25 percent alcohol, as shown by many tests, is the upper limit at which the blended fuel will vaporize satisfactorily in an unmodified carburetor.

England

England's early situation was slightly different. With her global oil companies firmly established, the Imperial confidence was high enough to shun legal measures until Britain entered World War II.

In the mid-1930s, two alcohol fuels were available in England. Cities Service Co. produced "Koolmotor," a blend of 16 percent ethyl alcohol and 84 percent gasoline. Cleveland Petrol & Distillers Co., Ltd., produced "Cleveland-Discol," a racing blend composed of about 79 percent ethyl alcohol, 9 percent acetone, and 10 percent gasoline.

Other European Pre-war Measures

Most nations on the Continent lacked the British confidence (and with good reason) and were cutting gasoline consumption years before their roles in the war-to-be were settled.

Beginning in 1931, the minister of finance in Austria had the authority to compel the addition of alcohol to gasoline when the wholesale price of the former was lower than the latter. A top limit of 25 percent alcohol was set, and the law required domestic alcohol. The modest result of this temperate law was the production of 6,786 hectoliters (180,000 American gallons) of alcohol in 1931-32.

Yugoslavia, in 1932, made it illegal to drive with any mixture other than 25 percent alcohol and 75 percent gasoline, and Poland compelled fuel dealers to add from 6 to 12 percent alcohol to gasoline.

Czechoslovakia, with a higher degree of industrialization, and a firmer attachment to the private vehicle than her neighbors, began mandatory use of alcohol fuel in 1928 with a blend called "Dynalkol," consisting of 50 percent alcohol, 30 percent gasoline and 20 percent benzol. Annual consumption was about 2½ million U. S. gallons in 1932, when all refined imported fuels were required to be blended with

20-30 percent alcohol, consisting of 95 percent ethyl and 5 percent methyl alcohols. The latter was made by wood distillation.

Secondary goals of the law were to increase alcohol production and potato farming. Seven wood-distillation plants were producing 35,000 hectoliters (924,700 U. S. gallons) of methanol per year.

In 1931 Italy had a law that called for 25 percent ethyl alcohol being mixed with all imported gasoline. Latvia passed a similar law, with purchase of alcohol being through a state alcohol monoply. Hungary's regulations provided for a 20 percent alcohol mixture with gasoline in a blend called "Motalko." Diesels and farm vehicles were given exemptions from using this.

Scandinavia

Sweden and other Scandinavian countries had limited highway systems and no crude oil production in the 1930s; they thus used little motor fuel compared with today's consumption. Sweden did offer various blends, called "Lattbentyl," of which 170,000 barrels were used in 1931. The Swedish people made alcohol from waste sulphate pulp liquor. The important part that the process could play in pollution control in the American paper industry is now being recognized.

Africa

South Africa was producing blends as early as World War I. During that war the Union of South Africa manufactured and sold "Natolite," a mixture of 60 percent ethyl alcohol and 40 percent ethyl ether, both sugar products. At the end of the war the components changed to 50 percent ethyl alcohol and 50 percent gasoline.

South America

Some South American nations, too, moved toward blends in the between-wars period. Argentina had enough petroleum to forestall such legislation, despite agricultural pressures to add alcohol made from grain, grapes and sugar cane.

Brazil's production of alcohol in 1932 reached about 16 million gallons, and she had no production of gasoline. Despite this imbalance, little effort was made to force regulations upon the affluent motoring minorities. In Pernambuco, two interesting blends were sold. One was a blend of 70 percent alcohol and 30 percent ether, and the other, sold in Rio de Janeiro and Nictheroy and called "Gasalco," was a blend of 82 percent alcohol and 18 percent gasoline.

Ether was needed, instead of an integral heater, to enable the engine to start without pre-heating either the intake air or the fuel. Alcohol, when used as a pure unblended fuel, does not readily vaporize in the standard carburetor, but when heated adequately, it becomes an ideal, cool-burning fuel. It is doubtful if the rich Brazilians, in their European sports cars and limousines, took advantage of the high-octane rating afforded by this racing fuel by increasing the compression ratios and advancing the spark timing.

Uruguay, Chile and Salvador, with state monopolies for fuel, had alcohol-additive laws similar to those in Europe.

Alcohol Fuel
in the United States

While nations in Europe and South America achieved varied success with alcohol-gasoline blends in that between-wars era, the achievements in the United States were minor, localized and short-lived.

Alcohol as a lamp fuel had a brief history in this country. Through the first third of the nineteenth century, whale oil was the preferred fuel for domestic lighting. With the increasing productivity of farms, grain was plentiful enough to provide alcohol as a lamp fuel, and the odorous animal and fish oils were replaced. The change emphasized three important qualities of alcohol fuel: It is odorless, nontoxic when burning, and the spilled fuel, if ignited, can be extinguished with water.

Then petroleum was discovered in Pennsylvania in 1859, and kerosene, because of its very low price, displaced alcohol as a lamp fuel, despite the odor of the new oil.

Dual Fuel System, 1900-1920

The early history of use of alcohol in the automobile in this country indicates that several early American automobiles were equipped with a carburetor heater and dual piping for alcohol and gasoline, thus providing complete interchangeability in fuels. The sectional drawing (illustration) from a maintenance text of 1911 shows a flexible fuel system, as well as illustrating the options in auxiliary equipment available to customers. Alcohol fuel equipment and instructions appear frequently in the texts, leading the reader to conclude that the fuel must have been accepted as equal to gasoline in most markets.

In these first two decades of the car, one of the principal advantages of alcohol fuel, its compatability with gasoline, was not utilized. Gasoline was used to start the car, and pure ethyl alcohol was used to run it, when the induction system became warm enough to vaporize the alcohol.

Now it appears more prudent to blend alcohol with gasoline, extending supplies of the latter and gaining, together, the merits of each fuel used separately. At some future date, when oil runs out, we may return to the pure alcohol vehicles of the past.

Tried After War

The first post-war wave of alcohol fuel blending came in 1922-23 when large stocks of World War I alcohol, made for munitions production, became surplus at low prices in the Baltimore area. Post-war motoring demands had pushed up the price of gasoline, and production facilities lagged behind. One of the largest oil companies took advantage of this temporary disruption to market a blend of the two fuels.

The results were unhappy. Alcohol absorbed water in the dealers' tanks, and led to separation, plugged fuel lines and erratic performance in customers' cars.

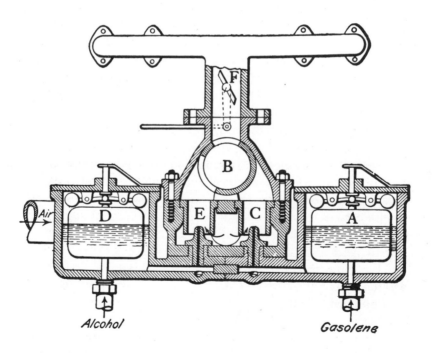

A double float-feed carburetor for alcohol and gasoline. The engine was started on gasoline from chamber A, and when the engine was hot, the rotary valve B was turned by the driver to use alcohol fuel from chamber D, drawn through jet E. F is throttle valve.

The Conflict with Gasoline

Although the causes of this fiasco were obviously avoidable, factors other than technical incompetence altered the future of alcohol. Among them was progress with gasoline. After 1923, new oil finds, the general expansion of the cracking process increasing the yield of gasoline and the use of tetraethyl lead as an anti-knock agent (replacing alcohol at one-tenth the cost) all lowered the price of gasoline. For about a decade alcohol fuels nearly disappeared, and were used only by the race drivers in their cars on state fair tracks and at Indianapolis.

Then came 1933 and the depths of the Depression with its farm surpluses and cheap labor to put otherwise idle distilleries into service. Idealistic and ecologically alert farmer groups encouraged government support of farm-produced fuel projects. They were successful in several states. First the Iowa Legislature proposed a law requiring the blending of alcohol with gasoline fuel. Then Nebraska and South Dakota removed state fuel taxes on the alcohol portion of blends.

But there were more failures than successes. Twenty bills in Congress and 31 in state legislatures failed to pass. Despite the low prices of grain, alcohol cost more than gasoline, and the ecologists, who wanted to spread organic fertilizer made from spent mash over the fields, began to give up. Politicians who had advocated the farm subsidies inherent in alcohol fuel production turned their attention to more direct methods of cash subsidy and support.

Private Efforts

There were efforts outside of government circles. In 1936 the Chemical Foundation, a Delaware corporation then thriving on impounded German patents and an enthusiastic agrarian reform philosophy, set up a distillery at Atchison, Kansas, to make "power alcohol" from farm surpluses. Atchison Agrol bought molasses, corn at 28 cents a bushel, and sold blended fuel for 25 cents a gallon. The plant closed after 2½ years of unprofitable operation.

Many reasons for the failure were advanced: The equipment was worn out at the start; the price Agrol paid for gasoline was too high and the federal Surplus Commodities Corporation failed to cooperate by selling grain at a price below the market.

A more ambitious experiment was tried by Herman Willkie, brother of the 1940 presidential candidate Wendell Willkie, and an officer of the Joseph E. Seagram and Sons Corporation. Willkie initiated research projects in his company and in the Indiana Farm Bureau. He tested tractors using alcohol blends, analyzed and improved processes of fermentation and distillation of alcohol fuel, and published a book, *Food for Thought.*

Willkie claimed to have research farms where sweet potato production reached 500 bushels per acre, releasing 500 gallons of alcohol. Some of the nitrogen removed from the soil by this intensive farming was returned by spreading the by-product mash as fertilizer.

Two requirements of Willkie's manifesto could be applied now to the agenda of the advocates of methanol utilization: (1) "A simple and portable continuous-process operation for conversion of the crops into alcohol" [substitute "organic wastes" for crops, today], and (2) "Internal combustion engines that will operate efficiently on [pure] alcohol."

Portable Distillery

For the first requirement Seagram drew up plans for a five-car railroad distillery, to operate continuously, with grain-to-alcohol conversion time of about eight hours. Other crops could be taken in, with ease. Willkie reported experimental use of wood wastes to produce fuel at 20 cents a gallon.

The project for the second requirement materialized modestly in a few prototype tractors by International Harvester, modified to high compression, for use in the Philippines. Another demonstration project, a Chrysler-engined tractor, was modified and tested for Seagram at the University of Kentucky.

Willkie's trainees at Seagram, from many of the "have-not" nations, particularly Latin America and India, were encouraged to return to their native lands with the technologies of advanced farming, natural (by-product) fertilizers, and farm-fueled tractors. However, famines, and the economic distresses of these countries, resulting directly from fertilizer shortages and energy pricing, all related to petroleum imports, rang a sad note of defeat to Willkie's high aspirations.

Agricultural alcohol motor fuel, although it never lost its sound basis in ecology, clearly became uneconomic in the 1940's. Today, with petroleum prices ascending, economics and general ecological pressures point the way for alcohol fuel's revival in the form of methanol. It would not be surprising if competitive pricing of alternative gaseous motor fuels drove us, literally, back to the beginnings of the internal combustion engine, the product of an era when coal gas was the preferred fuel for industry, cooking, lighting and small commercial power plants.

Origins of the Gasogen

The idea of harnessing blast furnace gas seems to have occurred to M. Lebon about 1800. He had invented illuminating gas and suggested that it might have been used a century earlier by those precursors of the internal combustion engine (Papin, Huygens, etc.) who dared use gunpowder as an engine fuel.

Lebon's idea lay unused until the development of Jean J. E. Lenoir's engine in 1860. But it was on other minds, not the least facile of which was Beau de Rochas's. He described the "four-stroke" cycle in 1862. Had the French been more aggressive, and had this scientist's name been shorter, the cycle might not have been called the Otto cycle.

Several experimenters contributed to the development of the modern "gazogene" in the decades prior to the important techniques of liquid fuel carburetting by Otto, Benz and Maybach, and the subsequent successful launching of the automobile on its errant course. Alfred Wilson and Emerson Dowson developed furnaces in which air and steam were injected into the combustion zone. The gas, known to the French as "gaz pauvre" (poor), had a low heat value because of the dilution of the principal combustible, carbon monoxide, with carbon dioxide and nitrogen. However poor it may have been, it was the first gas — a real gas, not the liquid we casually call gas — to feed the very first internal combustion engines, made by Jean Lenoir in Paris. It was also the first fuel to be manufactured from raw materials on board the moving vehicle, in a "gazogene portatif," at a later but uncelebrated date.

Lenoir's Drawing, 1860

The automobile of Jean Lenoir has been the subject of extensive historical speculation. It may not have existed in 1859 except in the minds of Lenoir and his overzealous press agent and fund-raiser, Gautier, and in Lenoir's patent application of that year. However, a drawing published in *Le Monde Illustre* in 1860, showing a high wagon, a one-cylinder horizontal engine, and a riveted gas pressure tank, established certain facts about the automotive art of the time — or the lack of it.

Although historians point out that the car could not have run because it lacked a clutch and other modern essentials, these questionable details are submerged by the well-documented fact that compressed coal gas was the fuel of the hour. The company formed by Lenoir's competitor, Hugon, was named "La Compagnie du Gaz Portatif" (Portable Gas Company). This gas was produced in a retort similar to those that continued to operate in some communities, where coal was economical, well into the twentieth century.

An Engine in Action, 1861

In 1861, Lenoir had the opportunity to install one of his engines in a launch being built by another inventor in the gas business, Lasslo Chandor. Gas for illumination was becoming popular, but if one's house were beyond the gas mains, what could be done to keep up with the urban fashions? Chandor Gas was distilled from naphtha and turpentine, and, presumably it was dispensed as a liquid, to be re-evaporated at the gas lamp. Although it was a clumsy way to provide gaseous fuel to a mobile engine, the hot tube probably used in this instance antedated the liquid carburetor, and might have made an automobile possible. Chandor and Lenoir certainly share honors for designing the first internal combustion motor boat. There is evidence that it navigated the Seine, but no data on performance are available.

Again traversing the thin ice of historical priorities, it may be claimed that Chandor gas was the first instance of the blending of a fossil fuel (naphtha) with a renewable fuel (methyl spirits and resins from wood, or turpentine).

There was no novelty in the production of city gas in the works of Paris and other capitals. A retort with a trap door and funnel at the top was the basic furnace; the result was piped into scrubbers, filters, tar separators, and so on.

The Portable "Gazogene," 1914

In the form of the portable "gazogene," the gas producer made its reappearance only in 1914. The prototypes for the portable equipment were found in the well-tested stationary gas producers of small size that made their appearance at the expositions of 1889 and 1900 in Paris. Instead of producing gas for storage in a tank, called a "gasometer," these more

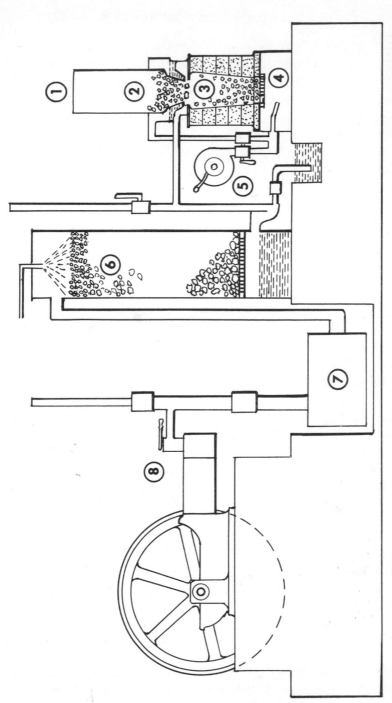

Where city gas was not available to run small industrial power plants, this self-contained suction gas producer and internal combustion engine was used. The fuel was coal. (1) filler cap, (2) coal magazine, (3) combustion, (4) ash pit, (5) starting fan, (6) scrubber, (7) dry filter, (8) throttle.

modern units were "direct induction." Professor Arbos, of Barcelona, was credited with this idea in a paper of 1862. It will be seen that the output of the generator will be in proportion, more or less, to the engine's speed, since the engine provides suction to maintain the fire in the generator.

The gasogen eliminated the bulky gasometer, with its water seals, and it produced gas, under slight vacuum, as was required by the engine (see illustration). Prior to this discovery, the injected or blown gas producers used petroleum, which may seem strange to us. However, oil was easily vaporized and metered into the furnace at a constant rate. Remember that the carburetor had not been perfected for use of volatile liquid fuel in an engine.

The First
Portable Gas Generator

"Poor gas" gained a foothold as an internal combustion engine fuel, in the typical stationary installation, and the solid forms of carbon, particularly charcoal and coke, were found to be more reactive than oil in a small retort, such as might be carried on a horseless carriage at a later date. The first truly portable vehicular gas generator to be installed in the United States was in a boat, the launch "Gloria," in 1908. Fifty-four feet overall, she was fitted with a four-cylinder engine and a coal-fired gas generator.

Fishermen and other commercial boat owners, converting their craft over from steam to the new gasoline and oil engines, felt the pinch of high fuel prices about 1910. This increase was reflected by the production in Boston of a line of gas producers for marine use. The Nelson Blower and Furnace Company, with engineer A. L. Galusha, equipped new and old boats up to 175 horsepower with money-saving gas generators until oil and engine prices dropped after World War I.

Economic Units

Although the Galusha gas systems appear bulky and heavy by modern standards of marine power plants, the owners judged them by comparison with the much heavier steam boilers and engines which were replaced, and their economy was evident. A construction launch of 50 horsepower, for example, cost $1.80 per hour to operate on gasoline in 1914 (at 12 cents per gallon). If the system were converted to pea anthracite coal (at $8.25 per ton), hourly cost was about 20 cents or one-ninth the gasoline cost. Now, with the cost of the two fuels elevated by a factor of five, the ratio remains about the same.

The First
Automotive Gas Generator

By a trick of fate, the romantic city of Casablanca became the site of the launching of that most unexciting, cumbersome, automotive anachronism of the first half of the twentieth century — the self-motivated gas generator. Under the sponsorship of the Automobile Club of Morocco, five trucks and tractors took part in a series of tests for farm vehicles during the war of 1914-1918. From this historic occasion until 1936, there are records of many efforts by the French Ministry of War and by other bureaus of research and invention to popularize the "gazogene" in a country that had no liquid fuel of its own, but whose practical people, obsessed with tourism, deluxe motor cars and racing, took a dim view of the cumbersome and temperamental "ersatz" gasoline machine.

The rallies went on for another decade in France. In 1925 there were 17 vehicles, French and Belgian. The road test, over a route of 2,100 kilometers, was followed by bench tests of six hours for each vehicle. The period was the peak of "gazogene" activity, with 25 types on the market.

Despite the stimulation of rallies sponsored by the French army and the agricultural bureaus, there were only about 2,000 "gazogene"-equipped trucks in use in 1938 — almost on the eve of World War II. There were very few innovations in French equipment in this final period, and, of course, none during the Occupation.

Problems During World War II

War in Europe brought severe belt-tightening, if not total curtailment, of all gasoline use by noncombatants. In 1940, gasoline prices were up to $1.03 per gallon in Italy, then a neutral. Alcohol was blended with petroleum fuel to extend it, and gas generators appeared in large numbers in every country in Europe. Germany's restrictions were the tightest: no civilian driving, no new automobiles, no uncontrolled use of gasoline. However, a synthetic fuel industry was very rapidly developed, using coal as the raw material. Even far-away New Zealand felt the pinch, and gasoline was rationed at 8 to 12 gallons monthly to civilians.

Britain

In Britain, where the domestic gas industry first flourished in the nineteenth century, the use of this commodity as a substitute motor fuel was the obvious answer for a coal-run country without oil. Gigantic gas bags, at atmospheric pressure, appeared atop buses and cars. They contained the equivalent of a half-gallon of gasoline, and they needed filling at frequent intervals. Trucks and buses often towed a trailer with a gas generator to retain adequate space for payload.

France

Although French reactions before the war were apathetic to the wholesale conversion of commercial and military vehicles to solid-fueled gas generators, their production of them in 1941 escalated remarkably. About 20 percent of the normal motor fuel needs were met by substitute fuels in that year. The Automobile Club of France announced that there were 60,000 charcoal-burning cars in operation and that 40,000 more were in production. Forestry reserves could support 300,000 vehicles thus equipped, it was estimated. Where youth camps replaced military service, charcoal production was the chief occupation, and in 1942, a total of 36,000 tons of charcoal per month was being produced.

Ethyl alcohol production for motor fuel blends was 10,000 tons that year, and this was expanded later by increased planting of alcohol-producing crops, notably sugar cane and beets.

A Maintenance System — Brazil

Brazil, with no important oil reserves of its own, anticipated its war-time problems by government encouragement of charcoal production and government-sponsored design of "gasogenios." In Sao Paulo, a commission of 60 was established. It trained 1200 servicemen for maintenance of the equipment. This wisdom paid off and made it possible to operate transportation into the interior, where petroleum fuel would not be a factor for years, or decades.

Scandinavia

Scandinavia, with its forests and its insignificant fossil fuel reserves, developed some excellent gas generators. The Svedlund System, made in Sweden, included a small boiler that injected steam with the air. The water thus introduced into the reaction zone was dissociated in hydrogen and oxygen, the latter combining with the incandescent fuel as CO. However, the variables in the process are too many in a small unit and the actual function of the steam was to reduce the temperature of the fire bed so that fusion and clinkers were avoided. Later models of the Svedlund System omitted the steam injection.

Methane Production

Sewage gas, containing enough methane to power an engine, has been seldom used because of its lengthy gestation in the works, about 30 days. One plant in Erfurt, Germany, compressed it into metal cylinders to 200 atmospheres (about 3,000 psi), enabling cars to make between 150 and 250 kilometers (91 to 155 miles) on a filling. Coal, charcoal and wood chips, because of the convenience of self-service and low price, remained the mainstay of the gas producers' industry in most countries of Europe and in South America.

A Final Note

In reviewing historical strategies of the rest of the world for conservation of motor fuels, North Americans should note a major difference between past solutions and the present dilemma. Ethyl alcohol, the mainstay additive of the 1930-1940 era, was produced without heavy capital investments, in decentralized distilleries, and from traditional raw materials: grapes and sugar-rich crops.

At present, methanol production from synthesis gas is only a large-scale operation, although work is being done to design small-scale plants, particularly portable equipment for use in extending wood waste salvage.

6

The Future of Cars
(and Fuels)

In January 1975, during the annual meeting of the American Association for the Advancement of Science (AAAS), there was a two-day symposium on "The Future of Cars." I attended, hoping to learn some new facet of methanol utilization. I also wanted to compare my own estimates of the car of the future with the opinions of others. Also at the meeting were many prominent figures in the Environmental Protection Agency, the Department of Transportation, several officers of three of the four automobile manufacturers, and many engineers and experts in various aspects of our leading product, the automobile.

Most of the talk was about the present: the existing state of pollution and of emissions-control law enforcement. We listened to reports on the dozen ways of making engines cleaner (all were too expensive, too slow), and on the economics of electric vehicles, of steam cars, of Stirling-cycle engines, on mining platinum for catalytic converters, on the substitution of rapid transit for the private car, and other subjects thought by their authors to be germane to the cleanliness, cost and salability of the car. Every means of improving the product was considered, except its fuel. When I inquired, during a coffee break, about the absence of the fuel question from the agenda, I was told that it was a subject not worth discussing. What can you do about it? If you have read this text, you can imagine my consternation with the reply.

Broad Knowledge Lacking: Why?

With the increasing complexity and specialization of science and technology, it is not surprising to find experts in fields as closely related as fuel combustion and automotive engineering to be relatively uninformed of each others' special knowledge. Even less surprising is the average car owner's confusion over the relationships among fuel economy, emissions control devices, and the politics of oil pricing and taxation. Unfortunately, the vested interests who profit from the status quo are complacent about this confusion. The automobile makers and the oil industry would be most unlikely to admit that there is a fuel that would solve two problems — pollution and costly oil imports — at the same time.

Fiction often clarifies the truth: Anyone involved with problem-solving on a professional level should be acquainted with the fictional Peterkin family and their methods. Mrs. Peterkin had put salt in her coffee instead of sugar. Her husband Agamemnon said, let us ask the Chemist what to do. He came over and added potassium and tartaric acid. He tasted, then added phosphoric, oxalic, acetic acids, and many more. But Mrs. Peterkin still objected to the taste. The Chemist was dismissed and the Herb Woman summoned. She tried dill, catnip, tansy, pennyroyal, valerian, hop and snakeroot. Still no good. As a last resort, the boys were sent out to ask the advice of the Lady from Philadelphia. Although busy, she sent back a message, suggesting that "your mother make a fresh cup of coffee."

The manufacturers have spent millions making experimental gas turbine cars, steam cars and exotic engines. Government agencies and urban organizations have spent or will spend a billion dollars to attempt to clean up the air, to undo the damage of years of neglect. Car owners are paying dearly for catalytic converters that are suspect of emitting an acid pollution of their own, and they are paying a price for gasoline that will only go up. Billions (of decreasingly valuable dollars, to be sure) are going to the Middle East and Africa for oil and for platinum. To maintain the flow of this single potent energy source, we have even whispered of military methods.

Despite all the research, expenditure, and risk, the manufacturers agree that no really radical changes in the basic internal combustion-engined car will be made before 2000 A.D. They have a solid record of being right.

After 12 hours of verbal floatation, I was ready to agree. Seventy-five years on one product is habit forming to an industry. However, the fuel will be changing before that time, because we are running out of oil. The affluent Iranians are not the last to realize this fact, and they are acquiring nuclear power plants as fast as possible.

Extending Fuels

Nuclear plants are not about to be available for automotive use, and they would not be found in submarines or cargo ships without the military requirements that often replace economic judgments. However, there are other economical and practical ways in which present fuel supplies may be extended. There are ways to phase in unfamiliar fuels without disrupting the economy — at least not to any greater extent than it is at present.

We have seen that methanol and wood gas generators could extend petroleum supplies and reduce the volume of imports. Since the United States Geological Survey has recently (June, 1975) acknowledged that our natural gas and oil resources had been overestimated in the past, more analysts have begun to agree with M. King Hubbert's forecasts of 1956, upheld in the 1970 peak production of 9.6 million barrels of crude oil, which has dropped off since that date. Asked recently if the administration's goal of terminating oil imports by 1985 was realistic, Dr. Hubbert replied that it was not possible, either by increasing our own outputs or by developing new wells.

The Effects of Government Taxes

Although most Americans are only recently aware of government's efforts to control foreign exchange deficits, particularly with the oil-exporting countries, Europeans have had far more painful reasons for complaint since the early 1930's. Then, a sampling of vehicle taxes was:

County	Average private car Tax per hp*	Actual tax of a 50 hp Rolls-Royce
Austria	$11.80	$590
England	11.80	590
Italy	11.65	580
New York and Pa.	1.65	83

---

*This horsepower was computed from a formula established in the age of steam.

At that time, vehicle taxation in England brought in a revenue of $138,500,000; and fuel tax receipts were $176,120,000 for a total of $314,620,000. The average total taxation per vehicle was $145 per year.

Those levels of taxation, in addition to reducing fuel imports, produced a hidden beneficial result: They induced manufacturers of cars to reduce the displacement (size) of engines and to develop the high-speed, high-compression sports car engine and chassis that became famous in England and Europe long before Detroit made feeble efforts to gain a fraction of the imported car market. It is quite unlikely that American manufacturers will be in a similar situation or will voluntarily spark a useful revolution in design.

Little Study of Fuels

While high taxes and raised prices on petroleum fuels have reduced sales for the first time in history, there has been little study of use of synthetic fuels from renewable sources. The problem of air pollution has become so entwined with gasoline economy in the minds of the public that much reeducative material is needed. The new head of Exxon, the world's largest industrial company, Clifton C. Garvin, has put it neatly:

> "What the United States needs is to reach a consensus for an energy program . . . . The public hasn't really been convinced yet of a need for a national energy policy."

After passing the buck to the poorly informed public, Mr. Garvin suggests that the government could be useful in development of substitute fuels, "exotic fuel forms, such as shale, oil from coal and solar sources." When war threatened Europe in the 1930's, Germany did not find coal so exotic, and built, in a couple of years, a half-dozen plants to convert it into aviation gasoline. We have even better technology now, but we dally in confusion while politics and profits dominate the public will.

Paths of Action —
Methanol and Power Plants

Two parallel paths of constructive action appear in the search for a synthetic automotive fuel policy. The first is the development of methanol production, as described in chapter 3, both for blending and for future 100 percent methanol-fueled engines. Second is continued development of non-petroleum power plants of the large, stationary kind now generating electricity, with concern for by-product fuels for vehicles. The primary energy may be nuclear, solar, tidal, wind, water, or geothermal. The part-time, or off-peak-load by-product might be electric current for vehicle batteries, metal salts for conversion in the vehicle to hydrogen gas, or other safely and easily handled liquids or gases on which conventional vehicles will run.

Another way to run a vehicle on stored energy, rather than burning fuel in the car, is to energize a built-in flywheel at stations equipped with electric or other motors. This has been accomplished on a Swiss bus line, and might be adapted to other short-haul services, such as suburban shopping and school vehicles.

Research Possibilities

The first part of this effort needs no research programs and no "exotic" technologies. It has all been done before, but for a change in alcohols. The second part needs research, and incentives for vehicle development that are beyond the motivations of the car and fuel manufacturers, if not beyond their billion-dollar prospecting budgets.

Cellulose to Glucose

Although ethyl alcohol is man's second-oldest conversion product — sour milk was probably first — a new way of making it from hitherto-unyielding cellulose, from wood, from paper, and from other renewable materials was developed in 1974. An enzyme was created by Dr. Mary Mandels and John Kostick of the U. S. Army Natick Laboratories that would break down cellulose to glucose in a few hours. Glucose is a sugar that converts, with yeast, to alcohol. Biological conversions of this kind are vastly more efficient,

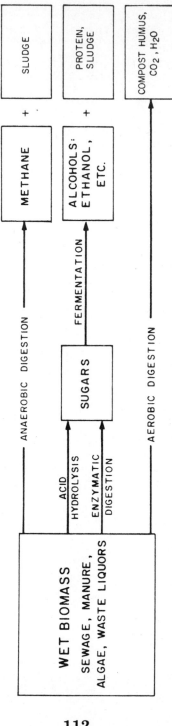

Conversion routes for wet biomass, such as sewage, to fuels.

in the heat they require, than the hydrogenation of coal or the synthesis of methanol.

A pilot plant to demonstrate the process is now in the design stage. A most promising source of cellulose for the long run (municipal rubbish is only about 50 percent cellulose) is manure, available from cattle feed lots in huge quantities. It is nearly all cellulose (see diagram).

Ethanol and "HydroFuel II"

The product of this new biological conversion process is mainly ethanol, with the impurities that render it unfit for drinking but make it an almost ideal additive for gasoline motor fuels. The superior mixing qualities at low temperatures and the higher heat content of ethanol over methanol are being utilized in tests and promotion of a proprietary product called "HydroFuel II," developed by United International Research, Inc.

This fuel consists of about 30 percent crude alcohol (including fusel oil, *t*-butanol, and up to 10 percent water) with about 70 percent gasoline, stabilized with an unstated amount of "Hydrelate," a synthetic fluid compatible with the others. While the test results of a 1973 unmodified automobile* are impressive, the vertical integration of biomass sources in a dispersed geographical pattern necessary to market the product is not likely for the near future.

Methanol for Diesels

The qualities of methanol fuel that make it ideal for spark-ignition gasoline engines — anti-knock, heat absorption,

---

*"Utilization of Hydrocarbon-Complex Formation in the Production of HydroFuel." A paper presented to the National Petroleum Refiners Association by Dr. Alfred R. Globus, president, United International Research, Inc., Hauppauge, N. Y., at the National Fuels and Lubricants Meeting, Sept. 11-12, 1975, Houston, Tex.

slow ignition — prevent its use in a diesel engine without modifications. However, a research project at MAN (Maschinenfabrik Augsburg-Nurnberg AG), aimed at reducing dependence on imported fuels, has also improved certain diesel operating characteristics. An ignition accelerator was added to the fuel, and preheated methanol was injected separately from oil fuel, later in the stroke. The experimenters reported "complete lack of soot in the exhaust gases" and an increase in torque without overloading. However, there are still problems with emissions of CO and HC.

Volkswagen Tests

Volkswagen has tested a gas turbine engine using methanol and found it to run with lower nitrogen oxide emissions than are produced on jet fuel. NO_x emissions over the legal limits was one of the problems arising with Chrysler's extensive testing of turbine cars although this was not a factor in the postponement of manufacturing. Rotary engines, those in which there are not reciprocating pistons, have potentials for fuel-saving and reduced emissions regardless of fuel used.

Continuous Internal Combustion

One system of great promise — theoretically, at least — investigated long ago for steam turbines is called *continuous internal combustion.* In a contemporary hypothetical case, a gas generator using solid or liquid fuel — either coal or methanol — is mated with a rotary engine or turbine. In such a closed system, gas quality can be controlled to refined limits; heat exchange is nearly perfect; there is no pre-ignition problem; all fuels are usable; there is minimal exhaust noise; and production costs should be low.

Volkswagen scientists have mentioned, in connection

with this "dream-car" system, a combustion catalyst that would dissociate methanol, providing hydrogen to the heat inputs.

Gas generators of "on-board" types deserve the revival of interest shown in a recent report from Germany. Their advantages, particularly with internal combustion engines, are clearly evident: liquid fuels can be "cracked" before entering the engine, hence eliminating nearly all emissions problems and all "skipping" or misfiring with lean mixtures. As these advantages become more important, along with use of coal and wood fuels as economic conditions indicate, the gasogen may come into the research lab for modernization and automation of its diverse functions.

The coming revolution in motor fuels, unlike revolutions of a political, technological, or economic nature, will occur gradually, but with absolute certainty, since it will stem from physical depletion of the supply of petroleum. The disinterest of the oil industry in prolonging supply by conservation was expressed candidly by the president of the American Petroleum Institute, Frank Ickard, on receiving criticism of high oil company profits in 1974: "Without profit, no business can grow. Without profit, consumer needs can't be met."

The API

Working quietly behind the scenes in Washington, lobbyists of the American Petroleum Institute try to see to it that its members are in no way threatened by legislation that might limit their profits or their freedom in perpetuating a nearly complete monopoly of world oil supplies. Their only failure, the inevitable termination of the depletion allowance, came about because most taxpayers could understand the iniquity of this loophole.

The API is dominated by the "Big Seven" companies — Exxon, Mobil, Standard Oil of California, Standard Oil of

Indiana, Texaco, Shell, and Gulf. In 1974, the institute's budget was over $15 million, supported by the dues of 265 corporate members and 7,000 individuals. The top eight members' dues are over $1 million each. Although the 1974 lobbying budget was only $200,000, the institute's role in influencing Congress is said to be far more significant than this sum would indicate.

Technical and statistical services of the API are maintained by a staff of about 330 employes, through publications, a speakers' bureau, and 11 lobbyists who supply congressmen with facts and figures not available from government sources.

The smaller oil companies, those not "integrated" in drilling, refining, distributing and selling their products, are organized in the Independent Petroleum Association of America.

The major oil companies will, according to Walter J. Levy, an economist, "continue to be the most important technical and marketing force in most of the world for a long time. Their technical competence and logistical services can only be replaced at a great risk to those who eliminate them." Serious though it sounds, this threat may be weakened by factors that outreach the great powers of the international companies.

The Challenge of a Methanol Industry

The strength of these companies depends, in varying degrees, on producing, transporting, refining, and marketing oil products. A domestic and regionally dispersed methanol industry, however, needs no prospecting, drilling, or pumping equipment. Raw materials — coal, wood, or rubbish — can dictate a close proximity to a synthesis plant, and no tanker fleets are necessary to bring the product to market. Two of the four critical activities of oil companies are thus obliterated. The synthesis plant for methanol is no more costly than the equivalent oil refinery, and it is difficult to imagine that the country's 200,000 filling stations would solidly refuse to sell anything but gasoline.

The use of methanol, introduced as a blend to municipal and large corporate fleets of cars and trucks, would naturally expand as the qualities of the fuel became evident to the private sector.

While coal must eventually replace oil as the basic fossil fuel of the world, this does not mean that there will be free competition for this product. The petroleum producers are already moving into coal mining, as they have into uranium. With ample advertising budgets, these companies will, in the near future, attempt to lull the public into the belief that all their research efforts, directed to improved strip mining and sulphurless combustion, are purely in the public interest.

7

What's a Driver to Do?

The American motorist stands at a curb beside his favorite possession, his automobile, and doesn't hide his concern. Gasoline prices have soared since those ominous days of the 1973 gasoline crisis. And while he still doesn't understand *why* there was a gasoline shortage in 1973, this motorist knows that the forces that locked the tanks and cut bulk gasoline deliveries then can do it again.

He asks the questions that all of us who drive cars frame in our minds: "What can I do?" And, "Can I avoid any future gasoline crisis — and still drive?" The answers aren't simple. But there is something he can do.

An Electric Car?

If his commuting distance is relatively short, 20 or so miles a day, and he has an extremely small family that prefers a short Sunday drive, perhaps an electric car would be a good buy for him. It will be turtle-slow rather than jackrabbit-jump at the lights, but he can ignore the gas stations, opened or closed.

If he is of a mechanical bent, he should read in this book about the gasogens. They're available — abroad — and they require fueling with wood, charcoal or coal, but they will provide him with safe, dependable power, and with no limit to distance, as with the electric car.

Conversion Kit?

Or, if he believes bottled gas is here to stay, there are conversion kits on the market today so that he gets his car fuel supply in a bulk tank. There's no major drawback to this system, unless it's the possibility of a gas as well as a gasoline shortage.

But our motorist, there on the curb, asks about the alcohol methanol.

He's heard it's a dependable fuel, could be a relatively cheap fuel, and could be made from many varieties of "trash," of which we have a good supply in this country. "If it's so good," he asks, "why can't I get methanol, and use it either as a blend in my present engine, or have a new engine that will burn straight methanol?"

The answer is that this may soon be possible, if you're willing to move to some other country, Germany perhaps, or one of the South American nations.

A Long Wait for Methanol

Methanol in this country for car fuel? Don't park your car in a metered zone while you wait for it, or you will push a lot of dimes into that meter. For the major oil companies are less than enthusiastic about any move toward methanol, and those in government seem peculiarly content to go along with the oil companies, even while loudly proclaiming their concern over energy shortages to any who are still listening.

Let's explore why this is so.

Twice in American history alcohol has been used as fuel, in limited geographical areas. Once was just after World War I, when there was a surplus of alcohol in the eastern part of the country. Gasoline was in short supply and relatively expensive, alcohol was plentiful and the price fell. While these conditions existed ethanol was used.

The second time was in the 1940s, in the Midwest, when alcohol was sold as fuel until the system providing it collapsed due to economic errors unrelated to the fuel itself.

Those two experiences, plus the widespread use of methanol for extended periods in other countries, have proved there are no technological reasons that bar the use of either major type of alcohol for automotive fuel.

Methanol Used Abroad

Other countries are taking major steps toward resumption of the use of methanol, knowing they must find some alternative to their dependence upon foreign oil.

The Volkswagen Corp. has begun a $1.5 million fleet test on a blend of 15 percent methanol in gasoline, with the test sponsored by the West German Government. A fleet of 45 cars is using fuel from six filling stations scattered through that country. The Germans remember the gasoline shortages of World War II and know they can produce methanol from brown coal at a reasonable price.

Volvo and the Swedish Government have established Svensk Metanolutveckling AB, a corporation to implement the use of alcohol blends and to develop production facilities. A 300-car fleet test is being started, using a blend of 16 percent methanol, 4 percent isobutyl alcohol (to prevent separation of the fuels at extremely low temperatures), and gasoline. Plans are underway to produce the methanol from the organic residue left from extracting uranium from lignite, and the additional supplies will be made from heavy oil residues.

Scientists in these countries and here are in virtual agreement about methanol: that it could lessen our dependence on imported oil, that it can be made from municipal and forest wastes (thus solving still another problem), that it would ease or eliminate auto engine pollution problems, and that it is an ideal fuel for the most efficient high-compression engines — the racing man's favorite.

Drawbacks of Gasoline

Gasoline, on the other hand, is far from ideal. It will hardly burn without knocking, unless doped with poisonous lead. Its exhaust is loaded with toxic gases and solids, controlled only by an expensive catalytic converter that emits sulphuric acid in quantities sufficient to dissolve the piping and muffler downstream from it in as little as 22,000 miles. Furthermore, continued use of petroleum products as fuel will remove this nonrenewable natural resource from its place as the essential material source of much of our technological production. Gasoline's virtue, in the past, was its low price.

Despite these comparisons of the two fuels, there is only talk and almost no action in the United States toward the use of methanol as an automotive fuel. The controversy is political and economic in character.

Opposition from Established Companies

The oil companies, with thinly disguised scorn, give methanol a bad report card. They are not content with the competitive situation today, with methanol selling at a higher price than gasoline. Instead they attempt to quiet any irreverent consumer curiosity about an alternative fuel, and they are anxious to localize the controversy to highly technical and often irrelevent theoretical matters, aimed above the average motorist's comprehension.

One shot was recently heard beyond the laboratory. *Science Magazine* (Nov. 21, 1975) reported that a fleet test for alcohol blends had been cancelled by MIT at its Energy Laboratory under circumstances that suggested pressure by corporations (Exxon and Ford) that support the research facility. Both MIT and the corporations denied that influence had invaded the high temples of university research.

This doesn't answer why the use of methanol seems to be opposed by American oil companies. and is ignored by United Stated government agencies set up for energy research and management. One would at first think that the oil companies would welcome a fuel that could be made from coal (much of which in this country is owned by the oil companies). This use, it would seem, would improve the performance of the gasoline with which it was blended, and eventually, would take the place of gasoline.

Since the top decisions of the oil companies are made in the privacy of their board rooms, one can only speculate on the reasons that have determined their policy against methanol.

Here are a few of the possibilities:

Competition. One is that manufacture of methanol and ethanol is relatively simple compared to the manufacture of gasoline. Furthermore the raw materials for methanol and ethanol, whether coal (sunshine stored for millions of years) or biomass (sunshine stored this year in cornstalks and hay or this century in wood) are widely distributed and accessible to many more groups than oil has been. Thus chemical companies, the government or new corporations could go into competition in supplying energy. The oil companies see themselves as the sole major purveyors of energy to consumers and have no interest in this technological opening to competition.

Capital. Another is that old invested capital is a universal barrier to the acceptance of new technology. When one builds a new plant, one convinces the bankers that this invested capital will be paying itself off in ten or twenty years, and one convinces one's stockholders that the profits will continue in perpetuity. The one threat to the argument is invention, which promises a cheaper or better product and makes obsolete previous plants. In this light, an oil company director would argue that profits are rising nicely now and let's get present equipment paid for by oil before we allow new technology and invention to make obsolete present plants and equipment. Since alcohols burn cleaner than gasoline there could be a hue and cry by the populace, demanding conversion to alcohol as soon as possible, thus saving the oil for its valuable chemicals, if the truth about methanol became widely known and accepted.

Exploration. Finally, the appearance of alcohols in the market could interfere with exploration for new oil. As long as there is no alternative to oil to keep us warm and moving in our cars, we are likely to pay any price for oil exploration, the greatest single expense of petroleum and gas production now. Of the wells completed in established fields in 1974, 37 percent were dry. In new fields, called "wildcats," 90 percent were abandoned, and only two percent proved to have more than a million barrels of oil.

As the odds against finding new oil go up, so does the cost of the gamble. In 1953 the average cost of drilling was $12.36 per foot. In 1973 this had risen to $20 per foot for inshore wells, and to $60 per foot for offshore wells. Adding to the instability of such high-risk ventures are the anxieties and ecological costs of drilling in the Santa Barbara channel and on the Grand Banks, and the hazards of supertankers on the shores of Alaska and the world's oceans.

Conversion to Methanol

Our concerned motorist has yet another question. If oil companies refuse to move toward conversion to methanol, why don't chemical companies take this step? Why aren't new corporations formed, solely to develop and put on the market this new fuel?

The answer today is a matter of simple economics. Despite the increase in gasoline prices and the probability that ultimately methanol will be cheaper, today it is more expensive than gasoline. Thus there is no incentive for other companies to set up the huge manufacturing and distribution system for a product that could not be priced competitively at this time.

It should be noted that the experiments in West Germany and Sweden are being underwritten in part by those governments. The United States government, operating on a philosophy of "what's good for the oil companies is good for the nation," has shown no similar inclination to produce such support.

The Cost

Before we become too pessimistic about replacing oil with synthetic fuels, let us assess the cost of doing this and how much it will hurt our pocketbooks. The problem is clear: We must build sufficient plant to manufacture about 100 billion gallons of liquid fuel per year, which is the rate of present gasoline consumption. And we must do this within the next 30 years, the period during which the remaining oil resources of this country are expected to be used up.

A number of recent estimates have put the cost of making synthetic fuel at about $1 billion for a plant capable of making one billion gallons, and we will have to build 100 equivalents of this — $100 billion — over the next 30 years. What an enormous sum of money, we say. And yet this is the cost that we have incurred in building our interstate highway system, and it hasn't been excessively painful. We presently pay a federal gasoline tax of 4 cents per gallon for those highways, and since we use 100 billion gallons of gasoline a year, this collects $4 billion a year. A similar tax could pay for the creation of the synthetic fuel plants that will continue to make our highway investment useful and would collect $120 billion over the 30-year period during which our oil is expected to run out.

Other Ways

We might raise this money in other ways. For example, as gasoline becomes more expensive, we could continue to sell synthetic fuel at a related price, with increasing profits. The profit from the early synthetic fuel plants could be used to build subsequent plants. Eventually the price of synthetic fuel would fall below that of natural gasoline and at that time no more gasoline would be produced. The remaining oil could be saved for valuable chemical production.

Like complacent political candidates, the oil companies are the incumbents, with a going concern, to state the situation mildly. At this time, there is no profit for them in a change of gasoline fuels.

This stand should not surprise us, since we have seen it before. In 1949 technological data were accumulated on the performance of alcohol fuels during the wartime years. The reports that came out of these data were strikingly different. In countries having ample oil supplies, the reports on performance were negative. But those countries without such resources reported good results from alcohol fuels.

Since the testing methods on which these reports were based left something to be desired, "one can scarcely avoid the conclusion that the results arrived at are those best suited to the political or economic aims of the country concerned, or of the industry which sponsored the research."[*]

Individual Action

We have seen that there are some steps the individual can take to avoid the discomfort of higher prices and eventual scarcity that are linked with dependence on the gasoline-powered automobile. But most of them are less than satisfactory, and particularly so when that individual knows something better is available.

He understands that this nation has the resources and the know-how to convert to one or more alternate fuels. Two

---

[*] "Alcohol — A fuel for Internal Combustion Engines." S. J. W. Pleeth. Chapman Hall, London, 1949.

forces are blocking the way to this goal, one that should be simple to attain for a nation that has unleashed the forces of the atom, and placed explorers on the moon. One is the oil industry, world-powerful and with understandably selfish reasons for wanting the status to remain quo. The other is his federal government. It is a government dedicated to the people, and not to its oil companies, yet a government that at best is bungling, bureaucratic and lethargic and at worst is led by many to whom the oil industry has supplied additional under-the-table income.

It is this government, huge and unresponsive as it is, that we must force into action if positive steps are to be taken in the crucial 30 years ahead.

Glossary
of Abbreviations,
Terms, and Symbols

AAAS   American Association for the
Advancement of Science.

API   American Petroleum Institute.

BTU   British thermal unit. Heat required to
raise 1 lb of water 1 degree Fahrenheit.
MBTU = 1,000 BTU. 1 = 0.25
calorie.

Cal   Calorie. Heat required to raise 1 gram of
water 1 degree centigrade.

C   Symbol for carbon.

CI   Compression ignition (engine). A diesel.

CO   Carbon monoxide, a combustible toxic
gas.

CO_2 Carbon dioxide, an incombustible inert
gas, found in the exhalation of all
animals.

CVCC   Controlled Vortex Combustion Chamber.

Dissociation   The process by which a chemical combination breaks up into simpler constituents, usually capable of recombining under other conditions. For example, methanol may dissociate thus: $CH_3OH \longrightarrow 2 H_2 + CO$, with the absorption of heat.

DOT   Department of Transportation.

EPA   Environmental Protection Agency.

HC   Hydrocarbon. Gases or liquids containing hydrogen and carbon.

IC   Internal Combustion.

IPAA   Independent Petroleum Association of America.

H   Hydrogen.

H_2O Water.

Hp   Horsepower. (1 Hp - 2545 BTU per hour.)

Kcal.   Kilogram-calorie, equal to 3.9 BTU (calories are more fattening).

LPG   Liquified petroleum gas.

MON   Motor octane number.

NO_x Nitrogen oxides, a pollutant of most IC engines.

O   Oxygen.

OPEC   Organization of Petroleum Exporting Countries.

RON   Research octane number.

Bibliography

"Automobile Aerodynamics, Form and Fashion." Karl Ludvigsen. *Automobile Quarterly,* vol. VI, no. 2 (1967).

Power Alcohol: History and Analysis. API publication, 1940. 58 pp.

Alcohol-Gasoline Studies at Massachusetts Institute of Technology. No. 3, API series, 1941.

Technical Characteristics of Alcohol-Gasoline Blends. American Petroleum Institute. Motor Fuel Facts Series, No. 1, 1938.

"Long Range Trends and the Future of the Automobile in America." Joseph F. Coates (Office of Technology Assessment, U. S. Congress). Paper presented to symposium on the Future of Cars, AAAS, Annual Meeting, Jan. 28, 1975.

"Self-Propelled Vehicles. A Practical Treatise on the Theory, Construction, Operation, Care and Management of all forms of Automobiles." James E. Homans, A.M. New York: Theo. Audel and Company, 1911. Esp. denatured alcohol, pp. 252, 253, 254., Fig. 180.

The Automobile — Energy and the Environment. A Technology Assessment of Advanced Automotive Propulsion Systems. Douglas G. Harvey and W. Robert Menchen. (NSF Contract, RANN, No. NSF-C674) Columbia, Md.: Hittman Associates, Inc., March, 1974.

"Now You Can Read Your Newspaper and Eat It, Too. Chemical Transforms Cellulose Waste into Food and Fuel." Bruce Myles, *Christian Science Monitor,* March 4, 1974.

Cost of Operating an Automobile. U. S. Department of Transportation, Federal Highway Administration, Highway Statistics Div. Washington, D. C.: Government Printing Office, April, 1972.

"EV (Electric Vehicle) Revival: A Special Report." *Machine Design,* October 17, 1974.

"The Energy Bazaar." Elizabeth Drew. (A Reporter at Large). *The New Yorker,* July 21, 1975.

The Energy Cartel: Who Runs the American Oil Industry. Norman Medvin. New York: Vintage Books, Random House, 1974.

"Energy and the Future." Allen L. Hammond, William D. Metz, and Thomas H. Maugh II. American Association for the Advancement of Science, 1973.

The Failure of U. S. Energy Policy. Richard B. Mancke. New York: Columbia University Press, 1974.

"Beating the Energy Pinch." Conrad Miller. *Motor Boating & Sailing,* March, 1974.

Man and Energy. Ubbelohde, A.R. Pelican A600,1963.

Food for Thought: Herman Frederick Willkie & Dr. Paul John Kolachov. Indianapolis, Ind.: Indiana Farm Bureau, Inc., c. 1942. [PPL No. B662.6. W 73 f.] A treatise on the utilization of farm products for producing farm motor fuel as a means of solving the agricultural problem.

On the Trail of New Fuels: Alternative Fuels for Motor Vehicles. Translated from German. Lawrence Livermore Laboratories, University of California, Livermore, Cal., June, 1975. Report of 15 institutions to West German Ministry for Research and Technology.

Fuel Cells: Power for Tomorrow. Daniel S. Halacy, Jr. Cleveland, Ohio: World Publishing Co., 1966.

"Potential for Motor Vehicle Fuel Economy Improvement." Report to the Congress, 24 October, 1974. USDOT and the USEPA.

"Biomass Energy Refineries for Production of Fuel and Fertilizer." Thomas B. Reed. Cambridge: Massachusetts Institute of Technology. Paper presented at Eighth Cellulose Conference, May, 1975.

Motor Fuel Economy of Europe. Dr. Gustav Egloff. API Series, No. 2, 1939.

Fuel: The Conquest of Man's Environment. Edwin Cecil Roberson and Roy Herbert. New York: Harper & Row, 1963. 128 pp.

The Principles of Motor Fuel Preparation and Application. Alfred W. Nash and Donald A. Howes. New York: John Wiley & Sons, Inc., 1935. Vol. I, "Alcohol Fuels," pp. 349-451; vol. II, "Motor Fuel Specifications."

"Old Fuel System May Be Answer to Costly Oil." Weston Farmer. *National Fisherman,* June-July, 1975.

Instruction Book for Government Utility Gas Producer, Marks VI and VII. London: HMSO, Ministry of War Transport, 1943.

"The Prospects for Gasoline Availability: 1974." A background paper prepared by the Congressional Research Service at the request of Henry M. Jackson, chairman, Committee on Interior and Insular Affairs, U. S. Senate. Serial No. 93-41 (92-76) Washington, D. C.: U. S. Government Printing Office, 1974.

A Text Book of Gas Manufacture. John Hornby. London: 1900.

Etude des Gazogenes Portatifs. G. Rouyer. Paris: Dunod, 1938.

"Is Hydrogen the Fuel of the Future?" Robert J. Trotter. *Science News,* vol. 102, July 15, 1972.

"A Sleeping Giant Awakes: The Invention of the Internal Combustion Engine." Robert F. Scott. *Automobile Quarterly,* vol. VI, No. 4, Spring, 1968. Page 410.

"Methanol as an Alternate Fuel." Reprints of 1974 Engineering Federation Conference, New England College, Henniker, N. H., July, 1974. (2 volumes)

"Improved Performance of Internal Combustion Engines using 5-20 percent Methanol." R. M. Lerner et al. Lincoln Laboratory, MIT., [n.d.].

"Sources and Methods for Methanol Production." T. B. Reed and R. M. Lerner. Lincoln Laboratory, MIT. [n.d.].

"Comparative Results on Methanol and Gasolene Fueled Passenger Cars." Herbert Heitland, Winfried Bernhardt and Wenpo Lee. R & D Division, Research Department., Volkswagenwerk AG, Wolfsburg, Germany, 1974.

"Methanol as a Gasoline Extender: A Critique." E. E. Wigg (senior research chemist, Exxon Research and Engineering Co., Linden, N. J.). *Science,* Nov. 29, 1974. Vol. 186, no. 4166.

"Methanol: A Versatile Fuel for Immediate Use." T. B. Reed and R. M. Lerner. *Science,* Dec. 28, 1973. Vol. 182, pp. 1299-1304.

The Peterkin Papers. Lucretia P. Hale. Boston: Houghton Mifflin Co., 1886.

Designing a Steam Car. William M. Davis. Stonington, Ct.: New Steam Age Publishing, 1942.

The Modern Steam Car and its Background. Thomas S. Derr. Newton, Mass.: First edition privately published by American Steam Automobile Co., 1932. Reprinted & Supplemented by Clymer Books, Los Angeles, California.

"Wood, Pulp ·and Paper, and People in the Northwest." J. Alfred Hall. Joint report of Northwest Pulp and Paper Association, Seattle, Wash., and Pacific Northwest Forest and Range Experiment Station Forest Service. Portland, Oregon: U. S. Department of Agriculture.

Index

Alcohol fuel, 2-3, 40; blends, 5, 6, 11, 28, 86-97, 104, 113, 117, 119-20, 123, 125; pure, 28, 92-93, 96-97

Aldehyde emissions, 26

Alterations for 100% methanol, 28-30

American Petroleum Institute (API), 11, 115-16

Automobile, designs for efficiency, 16-17; history, 14-16

Batteries, 74-78, 82-83, 111

Biomass, 3, 13, 35, 112-13

Carburetor shunt, 50

Catalytic converter, 27, 29, 79, 106, 107, 121

Charcoal, 4, 11, 32, 44-45, 48, 53, 56, 57, 101, 104, 105, 118

Citicars 73-74

City gas, 32, 44, 79-80, 99

Coal, 3, 4, 7, 8, 13, 17, 19, 30, 34, 36, 42-43, 47, 48, 52, 71, 73, 82, 87-88, 100, 103, 105, 110, 113, 115, 117, 118, 120, 122

Continuous internal combustion, 114-115

Cost comparisons, 19-20, 30-32, 38-40

Definition of methanol, 18-19

Elcar, 74

Electric car, 4, 13-14, 71-79, 106, 118; advantages, 71, 73-74; disadvantages, 72

Emissions 2, 3, 26-27, 30, 38, 40, 41, 47, 66, 69, 70, 79, 80, 83, 120; comparison with different fuels, 29; control of, 15-16, 27, 29, 59, 67, 73-74, 106, 107, 114, 115

Ethanol, 2, 3, 69, 89-90, 104, 105, 111-13, 119; compared with methanol, 18, 38, 113

External combustion engine, 43, 58, 69

Exxon tests, 26

Fertilizer, 36, 95, 97

Fossil fuels, 9-11, 73, 80, 99, 117

Fuel cells, 77-79, 81; for cars, 82

Fuel reserves, 8-9

Fuel tax, 109-10

Gasogen, 4, 7, 11, 13-14, 32, 39, 41-57, 71, 83, 115, 118; advantages of, 115; comparison of fuel consumption, 53, 102; costs, 56-57; design of, 46-52; disadvantages, 17, 42-43; origins, 97-105, 118

Gasoline, 5, 6, 10-13, 16, 17, 58, 79, 83-85, 86-91, 95, 96, 103, 108, 110, 113; analysis

133

(Gasoline, *continued*)
of, 53; compared with methanol, 23, 28, 30, 38-40; disadvantages, 2, 121; engine, 61, 66, 68, 84; shortage of, 86, 118; unleaded, 24, 26
Gas turbine engine, 70-71, 107, 114
Gunpowder as engine fuel, 97

Hydrogen, 5, 17, 32, 45, 69, 77, 79-83, 104; advantages, 80; disadvantages, 17, 82-83; production, 80-81

Internal combustion engine, 6, 7, 9-10, 43, 48, 58-60, 63, 69, 71, 79-80, 83, 96, 97, 98-99, 100-101, 108, 115

Lear steam turbine, 66
Lear, William, 63-64
Lenoir, Jean J. E., 9, 97-99

Manure, 113; digester, 20
Methane, 20; disadvantages, 17
Methanol, 3, 5, 6, 9, 67, 77, 79, 89, 96, 97, 113; advantages, 6-7, 13, 18-29, 30, 113, 116-17, 119, 120; compared with gasoline, 6-7, 19; cost, 19-20; disadvantages, 5, 17, 21-28, 30; production, 105
Methanol-gasoline 6, 13, 19, 20-30, 39-40, 79, 110, 119, 120; tests with, 20
Methanol power plants, 3, 33-34, 110-13
Motor fuels, conservation of, 86-105

Natural gas, 3, 8, 18, 32, 41, 82, 99, 108
Nuclear power, 31, 80, 82, 108, 111

Oil, 2, 3, 7-9, 11-13, 14, 32, 33, 73, 79, 87-88, 93, 95, 101, 108, 115-17, 120, 122-23, 125; companies, 6, 115-17, 119, 121-26; crisis, 12-13, 41, 82

Petroleum fuels, 9, 20, 41, 58, 69, 71, 82, 92, 93, 97, 101, 104, 115, 121; costs, 83
Production of methanol, 32-37; difficulties, 38

Raw materials of methanol, 3

Saab, 53, 55, 57, 67
Sewage gas, 105
Solar collectors, 3, 81-82
Stanley steamer, 60-62, 64
Steam, 32, 104; bus, 65-66; car, 4, 58-64, 66-67, 71, 73, 79, 84, 106-7; engine, 43, 58, 63, 66-68
Stirling cycle engine, 43, 58, 68-69, 79, 106
Supercharger, 52.
Svedlund, 50, 52-55, 57, 104-5
Synthetic fuels, 5, 6, 17, 87-88, 103, 110; economics, 124-25

Tuyeres, 47, 49

Vapor lock, 25-26
Volkswagen, 74, 120; tests, 3, 25, 26, 30-31, 114-15

Waste, municipal, 3, 7, 18, 34-37, 52, 71, 79, 88, 113, 116, 119, 120; wood, 7, 11, 13, 33, 105, 120
Water as fuel, 84-85
Wood, 3, 18, 33, 34, 42-43, 53, 82, 87, 96, 105, 115, 116, 118; gas generator, 56, 108